Concepts in Transition Metal Chemistry

Metals and Life and *Concepts in Transition Metal Chemistry* have been written as part of the Open University course S347 Metals and Life and are designed to work as stand-alone textbooks for readers studing either as part of an educational programme at another institution, or for self-directed study.

Details of this and other Open University courses can be obtained from the Student Registration and Enquiry Service, The Open University, PO Box 197, Milton Keynes MK7 6BJ, UK; tel: +44 (0)845 300 60 90; email: general-enquiries@open.ac.uk. Alternatively, you may wish to visit the Open University website at www.open.ac.uk, where you can learn more about the wide range of courses and packs offered at all levels by The Open University.

Concepts in Transition Metal Chemistry

Edited by Eleanor Crabb, Elaine Moore
and Lesley Smart

RSC Publishing

The Open University

Published by

Royal Society of Chemistry
Thomas Graham House
Science Park, Milton Road
Cambridge
CB4 0WF
United Kingdon

in association with

The Open University
Walton Hall, Milton Keynes
MK7 6AA
United Kingdom

Edited and designed by The Open University.

Typeset by The Open University.

Printed and bound in the United Kingdom by Halstan Printing Group, Amersham.

The paper used in this publication is procured from forests independently certified to the level of Forest Stewardship Council (FSC) principles and criteria. Chain of custody certification allows the tracing of this paper back to specific forest-management units (see www.fsc.org).

This book forms part of the Open University course S347 *Metals and life*. Details of this and other Open University courses can be obtained from the Student Registration and Enquiry Service, The Open University, PO Box 197, Milton Keynes MK7 6BJ, United Kingdom (tel. +44 (0)845 300 60 90, email general-enquiries@open.ac.uk).

www.open.ac.uk

British Library Cataloguing in Publication Data available on request

Library of Congress Cataloguing in Publication Data available on request

ISBN 978 1 84973 060 0 paperback

ISBN 978 1 84973 062 4 hardback

1.1

Preface

This book and the accompanying DVD-ROM aim to provide an accessible introduction to the chemistry of the transition metals and the main concepts in transition metal theory.

We begin, largely using interactive activities and video on the DVD, by introducing you to the chemistry of the first-row transition elements in various oxidation states, together with an exploration of the factors that influence the relative stability of the different oxidation states, in particular +2 and +3. We then move on to the field of coordination chemistry, complexes and ligands, considering also the stability of complexes. Later chapters look at theories of metal–ligand bonding, in particular the way models can be used to rationalise many of the chemical and physical properties of transition metals and their compounds, including their vibrant colours. Starting with the simple, yet powerful crystal-field approach, we then move on to a largely pictorial treatment of molecular orbital theory. (An appreciation of atomic and molecular orbitals as applied to the main-Group elements is assumed.) Throughout the text we have endeavoured to emphasise the practical relevance of the material by the inclusion of relevant experimental data and observations from everyday life, in particular in the field of bioinorganic chemistry.

At various points in the book, you will find 'boxed' material which provides background information or enrichment materials outside the main narrative of the text. Important terms appear in **bold** font in the text at the point where they are first defined, and these terms are also bold in the index.

Active engagement with the material throughout this book is encouraged by the use of embedded questions, indicated by a square (■), followed immediately by our suggested answer. The text is supplemented by multimedia activities on the accompanying DVD. This icon indicates when to undertake such an activity. (Please note that you will need to run the DVD on a computer as it contains interactive programs and is not designed to operate on a domestic DVD player connected to a television). In addition, further questions testing your understanding of the materials are included on the website associated with this book (indicated in the text by the icon). If you are studying this book as part of an Open University course you should visit the course website. If you are not reading this book in conjunction with an Open University course of study, further resources are available from the accompanying website by visiting www.rsc.org/metalsandlife.

We would like to thank the many people who helped with the production of this book. In addition to the principal authors, Kiki Warr contributed to the text. We would also like to thank David Johnson and Phil Butcher for creating the DVD which so clearly illustrates much of the chemistry of the transition metals.

We would also like to thank all those involved in the Open University production process, Margaret Careford for her careful word processing, Roger Courthold for transforming our rough sketches into colourful illustrations, Chris Hough for cover design and artwork, Hazel Carr, Yvonne Ashmore and Judith Pickering for managing the whole process and to our editor Rebecca Graham whose help in the editing of this book, is very gratefully acknowledged. We would like to thank our External Assessor, Professor Kieran Molloy, University of Bath and critical reader, Dr Ruth Durant, University of the West of Scotland, whose detailed comments have contributed to the structure and content of the book. Finally, we are delighted that this book is being published in association with RSC Publishing.

Contents

1 Introduction to the first-row transition elements

This book takes a close look at the transition elements, a family of elements whose chemical and physical properties contribute in a diverse way to many aspects of our lives, whether it be metabolic processes essential for our existence or playing a key role in the production of chemicals on the industrial scale. Take, for example, iron; this element is central to the transport and storage of oxygen in our bodies. It is also used as a catalyst in the large-scale production of ammonia and is the main component of stainless steel, one of the most important structural materials. Transition metals also form compounds whose colours span the spectrum – many gemstones, for example, owe their colour to transition metal compounds, as do many of the components of an artist's palette. But this is just a taster. In this chapter, you will explore some of the fundamental chemistry of these elements.

When studying this chapter, much of your time will be spent working through activities linked to the *Periodic Table* DVD. This text will guide you through the task, reviewing key points, expanding on others and reinforcing your understanding with a number of questions.

1.1 Introduction

To locate the transition elements, look at the long form of the Periodic Table shown in Figure 1.1. The elements in Period 4 that separate the main-Group elements on the left-hand side of the Table from the main-Group elements elements on the right-hand side, are the **first-row transition elements** or **first transition series**. They are the elements from scandium to zinc inclusive. The elements in Periods 5, 6 and 7 that lie beneath these are also transition elements. (The transition elements are sometimes referred to as the **'d-block' elements**.)

In this chapter we shall study some chemistry of the simple compounds and aqueous ions of the elements of the first transition series. Particular emphasis is placed on the general trends in properties that occur as we move from potassium and calcium, through the transition series that runs from scandium to zinc, to gallium, the succeeding **main-Group** or **typical element**.

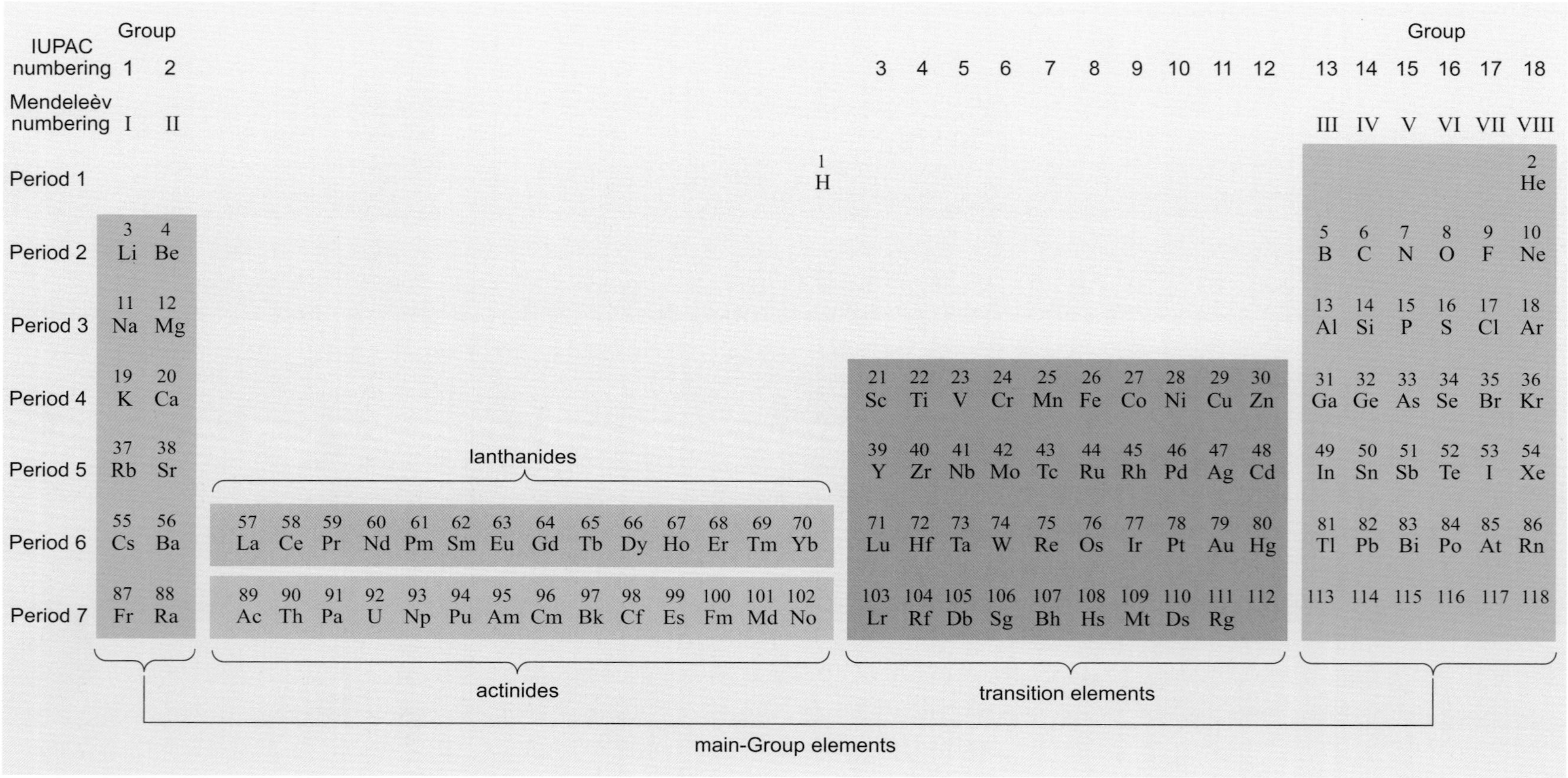

Figure 1.1 A long form of the Periodic Table.

Activity 1.1 The transition elements: preliminary concepts

At this point in the text, you should work through the four sections of the multimedia activity entitled *The transition elements: preliminary concepts*. (This is in 'Introducing the transition elements (d Block)'.) Details of how to access the activities are given at the back of this book.

The activity consists of four sections:

Section 1 is an introduction, which uses the chemistry of the typical elements (s- and p-blocks) to show how the replacement of highest valency by highest oxidation state provides a more appropriate insight into chemical periodicity. You may be familiar with the fact that the highest valencies of the elements can be used to detect chemical periodicity and construct periodic tables. However in this course, we will focus more on the concept of oxidation state, which has a special importance in transition metal chemistry.

Section 2 gives you the opportunity to look at samples of individual transition elements and to learn about some of their physical properties.

Section 3 covers the electronic configurations of transition metal atoms and their ions.

And finally, Section 4 explores the relationship between electronic configuration and oxidation state in transition element chemistry.

1.2 Electronic configurations of the transition elements

Let us consider the first transition series elements in Period 4. The electronic configuration of the free argon atom at the end of the previous row is $1s^22s^22p^63s^23p^6$. Moving now from potassium to zinc, electrons enter the 4s or 3d levels; for these elements, both the 4s and 3d electrons are **valence electrons**. In the potassium and calcium atoms, the valence electrons enter the 4s level, so the electronic configuration of calcium, for example, can be written $[Ar]4s^2$. At scandium the 3d level begins to fill. An appreciation of this process is pivotal to your understanding of transition metal chemistry, and the electronic configurations of the free atoms of the first-row transition elements and their dipositive ions are reproduced in Table 1.1.

You can see that there is a gradual filling of the 3d shell across the series. This filling is not quite regular, however; at chromium and copper the population of the 3d level is increased by the acquisition of one of the two 4s electrons. Apart from zinc and copper, the free atoms have incomplete 3d shells.

As you will see in later sections, in many ways, the chemistry of the transition elements is more easily related to the electronic configurations of free ions than to that of free atoms.

Table 1.1 Electronic configurations of the free atoms and dipositive ions of the first transition series.

Element	Free atom	Free M^{2+} ion	Element	Free atom	Free M^{2+} ion
Sc	$[Ar]3d^14s^2$	$[Ar]3d^1$	Fe	$[Ar]3d^64s^2$	$[Ar]3d^6$
Ti	$[Ar]3d^24s^2$	$[Ar]3d^2$	Co	$[Ar]3d^74s^2$	$[Ar]3d^7$
V	$[Ar]3d^34s^2$	$[Ar]3d^3$	Ni	$[Ar]3d^84s^2$	$[Ar]3d^8$
Cr	$[Ar]3d^54s^1$	$[Ar]3d^4$	Cu	$[Ar]3d^{10}4s^1$	$[Ar]3d^9$
Mn	$[Ar]3d^54s^2$	$[Ar]3d^5$	Zn	$[Ar]3d^{10}4s^2$	$[Ar]3d^{10}$

The relevant ions have a charge of +2 (+1 in copper) or more, and their configurations can be obtained by removing, *first*, the outer s electrons of the free atom and, *second*, the outer d electrons, until the total number of electrons removed is equal to the charge on the ion. For example, the electronic configurations of the dipositive ions obtained by applying this generalisation are shown in columns 3 and 6 of Table 1.1. Note that, unlike the free atoms, there is now a smooth progression in the 3d electron population.

■ The electronic configuration of the free chromium and cobalt atoms are $[Ar]3d^54s^1$ and $[Ar]3d^74s^2$. What are the configurations of the free ions Cr^{3+} and Co^{4+}?

□ On removing the s electrons first, and then the d electrons, we are left with Cr^{3+}: $[Ar]3d^3$ and Co^{4+}: $[Ar]3d^5$.

1.3 The elemental state

Figure 1.2 shows samples of chromium, manganese, iron and cobalt. These elements, and indeed all the transition elements, are metals. This generalisation suggests that the transition elements are more alike than the main-Group elements, which include both metals and non-metals. At room temperature, apart from manganese, the metals of the first transition series all have one of the three common metallic structures: body-centred cubic (*bcc*), cubic close-packed (*ccp*) or hexagonal close-packed (*hcp*), as shown in Figure 1.3a, b and c, respectively. (See also Box 1.1.)

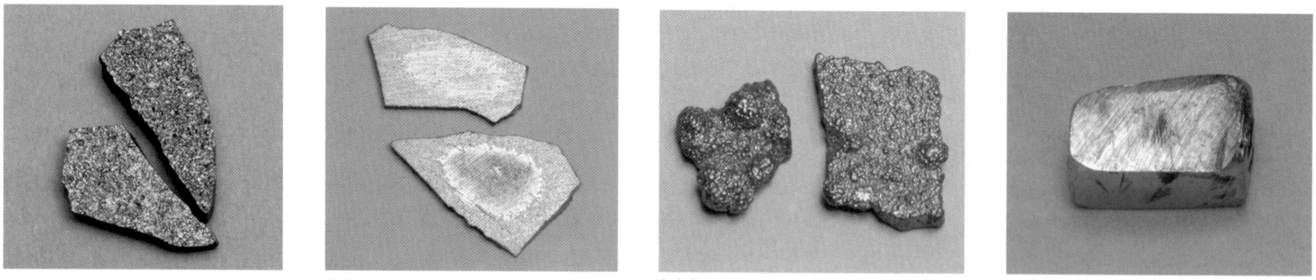

(a) chromium (b) manganese (c) iron (d) cobalt

Figure 1.2 Samples of (a) chromium, (b) manganese, (c) iron and (d) cobalt.

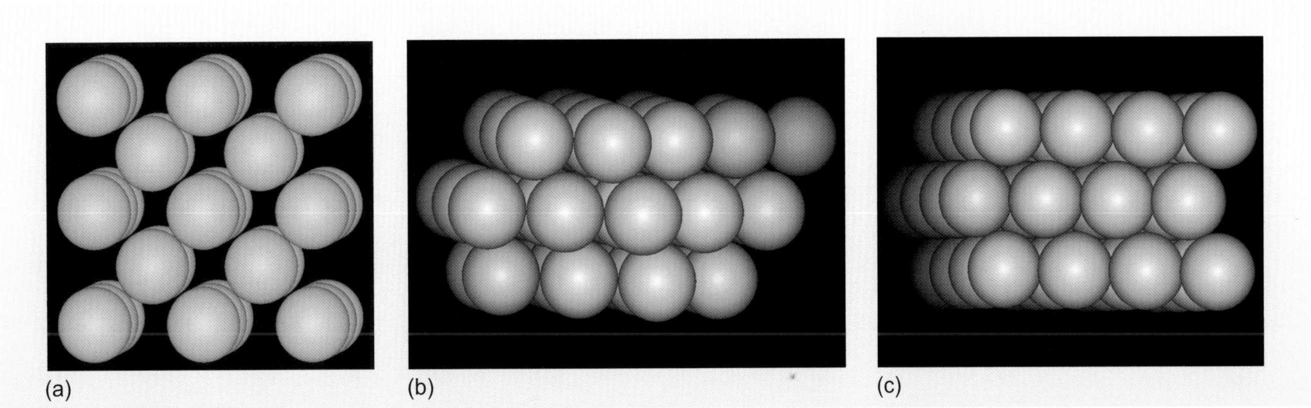

Figure 1.3 Computer representation (side view) of layers showing (a) *bcc*, (b) *ccp* and (c) *hcp* stacking.

Box 1.1 The three common metallic structures

If we consider the atoms in a metal as hard spheres, then the structure of the metal can be described by the three-dimensional packing of these spheres. In the most efficient form of packing there are minimal gaps between the spheres. This is known as close-packing.

Figure 1.4 shows close-packing of two layers of spheres, A and B.

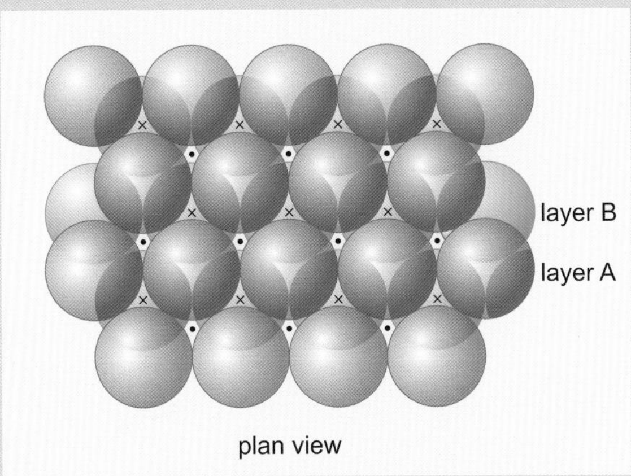

Figure 1.4 Plan view (from above) of two layers of close-packed spheres.

When we add another layer of spheres, there is a choice where they can go. In the first alternative, the spheres in layer C can be arranged so that they are directly over the spheres in layer A; they are placed over the tetrahedral holes in layer B (marked with crosses in Figure 1.4). This type of packing is known as hexagonal close-packing, *hcp* (shown in Figure 1.3c). The layers in a hexagonal close-packed structure can be designated as ABAB… . The alternative arrangement is for layer C to lie over the octahedral holes in layer B (marked with spots in Figure 1.4). This is known as cubic close-packing, *ccp*, and is designated ABCABC… (as in Figure 1.3b). This is also referred to as face-centred cubic, *fcc*. In both these forms of packing each sphere is surrounded by 12 others.

Other metals adopt less efficiently packed systems, such as the body-centred cubic (*bcc*) structure (Figure 1.3a) in which an atom is in the centre of a cube surrounded by eight equidistant atoms at the corners of the cube.

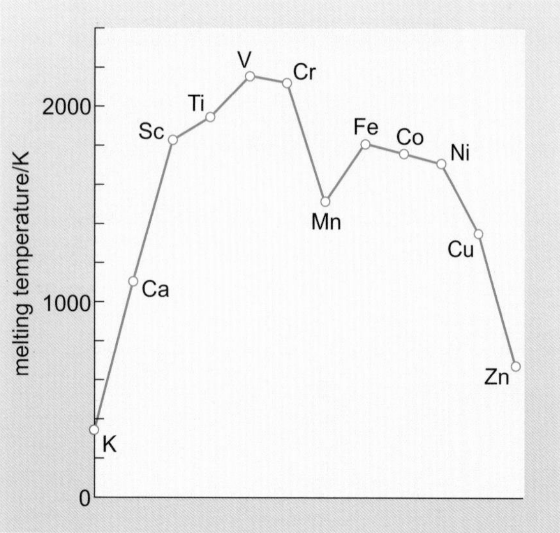

Figure 1.5 Melting temperatures of the elements from potassium to zinc inclusive.

The structures of all the first-row transition elements at room temperature are given in Table 1.2, and there are, as you may recall, similarities to those of some of the main-Group metals. For example, chromium and iron have the same structure as sodium; cobalt is isostructural with magnesium and copper with aluminium. Evidently the introduction of d electrons does not greatly influence the structure of metals. However, it does influence other important properties such as the melting temperatures and the boiling temperatures. The former are also given in Table 1.2. The melting temperatures of sodium, magnesium and aluminium are 371 K, 924 K and 933 K, respectively. You can see that, with the exception of zinc, all the first-row transition metals melt at least 400 K higher than aluminium. The transition metals are also harder than the typical metals. If the transition elements are harder and have higher melting temperatures (and boiling temperatures) than the typical elements, this can be attributed to the stronger forces that bind their atoms together. These can be crudely correlated with electronic configuration. In Figure 1.5 we have plotted the melting temperatures of the metals from potassium to zinc.

Table 1.2 Structures of the first-row transition elements at room temperature and their melting temperatures (m.t.).

Element	Structure	m.t./K	Element	Structure	m.t./K
Sc	*hcp*	1812	Fe	*bcc*	1808
Ti	*hcp*	1933	Co	*hcp*	1768
V	*bcc*	2163	Ni	*ccp*	1726
Cr	*bcc*	2130	Cu	*ccp*	1356
Mn	*	1517	Zn	*hcp*	693

* The structure of manganese is complex but the coordination number of 12 is large as in the common metal structures.

■ Which four elements in Figure 1.5 have the lowest melting temperatures?

□ Potassium, calcium, copper and zinc.

■ What do you notice about the electronic configuration of these four atoms?

- The 3d shell is either empty or full. They have no unpaired d electrons.

You can think of unpaired d electrons in the free atoms as electrons available for the formation of particularly strong electron-pair bonds. Note that between calcium and scandium, where a d electron first appears, there is a jump of nearly 700 K in the melting temperature. The presence of one or more unpaired d electrons thus leads to higher interatomic forces and, therefore, to high melting and boiling temperatures. Note that this simple argument cannot be made more precise by saying that the melting temperatures increase with the number of unpaired electrons in the ground-state configuration of the atom: for example, manganese, with five unpaired d electrons, has a lower melting temperature than titanium and nickel, which have only two. Nevertheless, even the most sophisticated theories suggest that strong bonds formed by d electrons are responsible for the particularly high melting and boiling temperatures of the transition metals.

In contrast to the **alkali metals** or **alkaline earth metals**, most transition metals are not readily attacked by air or water and this, together with their hardness and high melting temperatures, gives them great industrial importance. This is especially true of the four metals shown in Figure 1.2. Chromium, manganese and iron are among the more plentiful metals in the Earth's crust; after aluminium, iron is the most abundant metal and chromium and manganese are more common than copper or zinc. Iron is involved in many manufactured goods, and chromium and manganese play vital alloying roles in the steel industry: for example, stainless steels are alloys containing mainly iron with about 13% chromium and 0.5% manganese. Cobalt is a constituent of important specialist alloys and of magnetic materials with electronic applications.

1.4 Oxidation state +2

In this section, we shall examine an important example of the +2 oxidation state of the first-row transition elements: that of the dipositive aqueous ions. We shall use these ions to draw out some general characteristics of transition elements, most notably colour and the extent to which it can be related to electronic configuration. In doing this, occasional reference will be made to other elements and other oxidation states. In particular, potassium and calcium, the two elements that precede the first transition series and gallium, the succeeding main-Group element, will be included in our study.

Activity 1.2 Chemistry of the transition elements

You should now work through the first two sections of the multimedia activity entitled *Chemistry of the transition elements* in 'Introducing the transition elements (d Block)':

Section 1: Introduction

Section 2: Colour and d-electron configuration

1.4.1 Aqueous ions

On the DVD, we examined the dipositive aqueous ions M^{2+}(aq) of the first transition series and their chemistry. The last eight of the ten first-row transition elements form aqueous dipositive ions. Scandium and titanium do not.

Table 1.3 The known dipositive aqueous ions, their colours and electronic configurations for the first-row transition metals.

Aqueous ion	Colour	Free M^{2+} ion	Aqueous ion	Colour	Free M^{2+} ion
Sc^{2+}	(ion not known)	$[Ar]3d^1$	Fe^{2+}	pale green	$[Ar]3d^6$
Ti^{2+}	(ion not known)	$[Ar]3d^2$	Co^{2+}	pink	$[Ar]3d^7$
V^{2+}	lavender	$[Ar]3d^3$	Ni^{2+}	green	$[Ar]3d^8$
Cr^{2+}	sky-blue	$[Ar]3d^4$	Cu^{2+}	blue	$[Ar]3d^9$
Mn^{2+}	very pale pink	$[Ar]3d^5$	Zn^{2+}	colourless	$[Ar]3d^{10}$

The colours of the dipositive aqueous ions are shown in Table 1.3. You can see that the solutions of the transition metal ions are coloured (although that of Mn^{2+}(aq) is very pale), with the exception of Zn^{2+}(aq).

■ Can you see a link between whether a solution is coloured or not and the electronic configuration of the ion?

□ Only Zn^{2+}(aq), with a complete d sub-shell is colourless. Colour is often associated with partially filled d sub-shells, in particular it is frequently due to transitions between levels within the d sub-shell of a metal ion. We will learn more about the colour of transition metal ions in Chapters 5 and 6.

We will now consider an example of the transition elements in the same oxidation state of +2 but in a different environment; that of the dihalides.

1.4.2 Dihalides

Many first-row transition metals form difluorides, dichlorides, dibromides and diiodides (Table 1.4). In fact all dihalides of the first-row transition metals are known except for TiF_2, CuI_2 and the dihalides of scandium.

Table 1.4 The range of known dihalides of the first-row transition elements.

Element	Dihalide ion	Element	Dihalide ion
Sc	none known	Fe	all known
Ti	all known except TiF_2	Co	all known
V	all known	Ni	all known
Cr	all known	Cu	all known except CuI_2
Mn	all known	Zn	all known

■ Do you detect any relationship between this generalisation and the generalisation made earlier about the occurrence of dipositive aqueous ions?

□ In the dispositive aqueous ions, the elements for which such ions are unknown are scandium and titanium, which occur at the beginning of the transition series. Five of the six unknown dihalides are also compounds of these two elements.

There have been attempts to make these dihalides and dipositive ions, and the fact that they may well be unknown suggests they are unstable with respect to some decomposition reaction. We will explore this further in later sections.

1.4.3 Some reactions within the dipositive oxidation state

In reactions in which these dihalide compounds or ions are interconverted and in which there is no change in oxidation state, the first-row transition metals behave similarly. For example, if the dichlorides of vanadium, chromium, manganese, iron, cobalt, nickel, copper and zinc are added to de-aerated water under nitrogen, they all dissolve to form dipositive ions:

$$MCl_2(s) = M^{2+}(aq) + 2Cl^-(aq)$$

Aqueous solutions containing the dipositive ions also behave similarly in several reactions. The addition of sodium hydroxide and sodium carbonate solution in most cases precipitates insoluble hydroxides and carbonates, respectively, as follows.

$$M^{2+}(aq) + 2OH^-(aq) = M(OH)_2(s)$$

$$M^{2+}(aq) + CO_3^{2-}(aq) = MCO_3(s)$$

The solubilities of the hydroxides $M(OH)_2$ are similar, as are those of the carbonates.

(*Note*: for copper and zinc we have to make a slight stipulation, carbonate precipitates basic carbonates such as $2CuCO_3.Cu(OH)_2$ from solutions of $Cu^{2+}(aq)$; $Zn(OH)_2$ is amphoteric and dissolves in excess NaOH:

$$Zn(OH)_2(s) + 2OH^-(aq) = [Zn(OH)_4]^{2-}(aq).)$$

However, the key point is that similarities of the kind described above are not maintained when we turn to redox reactions in which there is a change in oxidation state of the transition metal species. As we shall find, the redox chemistry of the transition elements, in addition to being a colourful property, reveals great variety between the different metals. We will now look at some of the other oxidation states of the first transition series.

1.5 A survey of oxidation states greater than +2

In the next activity you will view a series of experiments on the DVD which will enable you to construct a chart showing the distribution of oxidation states for the elements potassium to zinc.

As the title of this section implies, we will mainly be concerned with oxidation states greater than +2. A major focus of the study will be an examination of the trend in the highest oxidation state of the elements and its relationship to the changing d-electron configuration of the metal.

Activity 1.3 A survey of higher oxidation states

You should now work through Section 4 of *Chemistry of the transition elements*.

Section 4: A survey of higher oxidation states.

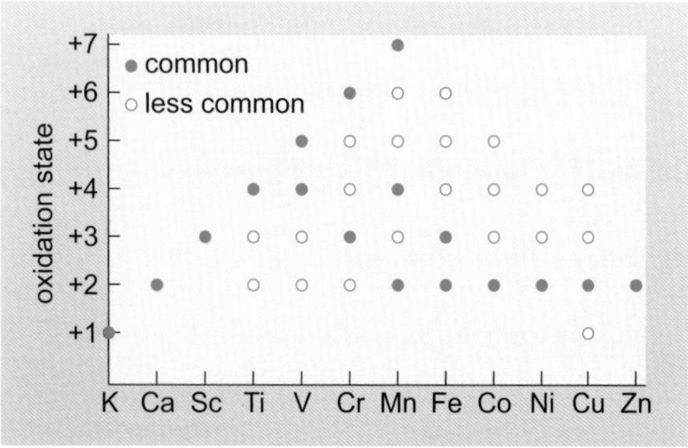

Figure 1.6 Oxidation-state patterns for the elements from potassium to zinc, compiled from a survey of aqueous monatomic ions, aqueous oxo-ions, binary halides and oxides and solids containing fluoride and oxide complexes.

Figure 1.6 summarises the oxidation states of the elements from potassium to zinc derived in Activity 1.3 above. It gives an accurate picture of the distribution of oxidation states +2 and above. The filled circles denote the most common oxidation states of each element – those often found on the laboratory shelf. Unlike the compounds of oxidation states marked by open circles, these compounds are usually considerably stable with respect to oxidation and reduction.

■ Can you see a pattern with respect to the highest oxidation state across the first transition series?

□ The highest oxidation state follows the number of outer valence electrons from potassium to manganese.

Beyond manganese, the highest oxidation state drops increasingly below the number of outer electrons as these electrons become more tightly bound by the increasing nuclear charge and it becomes impossible to synthesise compounds in which all the outer electrons are used in bonding. As the nuclear charge increases, 3d electrons become more tightly bound than 4s electrons. This reveals itself most notably in the chemistry of zinc whose compounds are confined to the +2 oxidation state as the ten 3d electrons have become so stabilised that they are, for all chemical purposes, part of the inner core of electrons, and not available for bonding. In many respects, therefore, zinc behaves like an alkaline earth metal.

1.6 A closer look at the +3 oxidation state

The redox chemistry we examined in the previous section introduced you to many new oxidation states having oxidation numbers greater than +2. Particular emphasis was placed upon the highest achievable oxidation state and the way in which it varied between potassium and zinc. However, in the course of the survey we also encountered the +3 oxidation state of the first transition series in the form of the aqueous ions, $M^{3+}(aq)$. The distribution of known ions of this kind is different from that of the dipositive ions, $M^{2+}(aq)$, which we studied in Section 1.4. We shall now examine it more closely.

1.6.1 Aqueous ions

Table 1.5 shows the full range of known tripositive aqueous ions of the first-row transition metals.

Table 1.5 The known tripositive aqueous ions and their colours for the first-row transition metals.

Aqueous ion	Colour	Aqueous ion	Colour
Sc^{3+}	colourless	Fe^{3+}	yellow-brown
Ti^{3+}	purple	Co^{3+}	blue
V^{3+}	green	Ni^{3+}	ion not known
Cr^{3+}	green-violet	Cu^{3+}	ion not known
Mn^{3+}	claret	Zn^{3+}	ion not known

■ Summarise concisely the extent to which the first-row transition elements form tripositive aqueous ions.

□ The first seven elements of the first transition series form tripositive aqueous ions; the tripositive ions of the last three members, $Ni^{3+}(aq)$, $Cu^{3+}(aq)$ and $Zn^{3+}(aq)$, are unknown.

■ In what way does this statement contrast with the statement made in Section 1.4.1 describing the range of known *dipositive* aqueous ions?

□ The missing tripositive ions occur at the end of the transition series but the unknown dipositive ions, $Sc^{2+}(aq)$ and $Ti^{2+}(aq)$, occur at the beginning.

Now consider the *relative* stabilities of dipositive and tripositive aqueous ions between scandium and zinc; suppose that the stabilities of the $M^{2+}(aq)$ and $M^{3+}(aq)$ ions with respect to each other determine the question of the existence or non-existence of $M^{2+}(aq)$ and $M^{3+}(aq)$ for each element.

■ What pattern emerges across the first transition series in terms of the relative stabilities of the dipositive and tripositive aqueous ions?

□ It appears that oxidation of $M^{2+}(aq)$ to $M^{3+}(aq)$ is so easy at the beginning of the series that $M^{2+}(aq)$ is unknown; on the other hand, it is very hard at the end of the series where $M^{3+}(aq)$ is unknown. Thus it looks as if the job of oxidising $M^{2+}(aq)$ to $M^{3+}(aq)$ becomes harder as we move from scandium to zinc.

Activity 1.4

We shall test the hypothesis that oxidation of M^{2+}(aq) to M^{3+}(aq) becomes harder from scandium to zinc further by looking at some more reactions of the dipositive and tripositive aqueous ions.

You should now work through the next section of *Chemistry of the transition elements*:

Section 5: Relative stabilities of the +2 and +3 aqueous ions.

Don't forget that there are questions on the companion website which you can use to test your understanding of the material covered in this chapter.

We will consider the difference in relative stabilities of the different oxidations states in Chapter 2.

2 Thermodynamic stability

At the end of the previous section, we made several observations regarding the reactions of the dipositive and tripositive ions of the first transition series, which suggested that the difficulty of oxidising M^{2+}(aq) to M^{3+}(aq) varied as follows:

$$\mathrm{Sc < Ti < V \sim Cr < Mn > Fe < Co < Ni < Cu < Zn}$$

In general, it becomes progressively harder to oxidise M^{2+}(aq) to M^{3+}(aq) as we move across the series, but there is an important exception in the form of a downward break between manganese and iron.

Although it is clearly possible to draw conclusions about the relative stabilities of M^{2+}(aq) and M^{3+}(aq) on the basis of simple test-tube reactions, we need to be careful, as we have not distinguished between thermodynamic and kinetic data.

We will now attempt to put the process on a more quantitative footing. The sequence above is supported by thermodynamic data for Reaction 2.1. The values of $\Delta G_m^\ominus$ (the standard molar Gibbs function change) for this reaction increase from scandium to zinc with the exception of a slight decrease between vanadium and chromium and a decrease between manganese and iron, as shown in Table 2.1.

$$\mathrm{M^{2+}(aq) + H^+(aq) = M^{3+}(aq) + \tfrac{1}{2}H_2(g)} \qquad (2.1)$$

Table 2.1 Values of $\Delta G_m^\ominus$ for Reaction 2.1 and of $E^\ominus(M^{3+}|M^{2+})$ for the metals of the first-row transition elements at 25 °C.

Metal	**$\Delta G_m^\ominus(2.1)^*$ / kJ mol^{-1}**	**$E^\ominus(M^{3+}\|M^{2+})^*$ / V**	**Metal**	**$\Delta G_m^\ominus(2.1)^*$ / kJ mol^{-1}**	**$E^\ominus(M^{3+}\|M^{2+})^*$ / V**
Sc	(−251)	(−2.6)	Fe	74	0.77
Ti	(−106)	(−1.1)	Co	187	1.94
V	−25	−0.26	Ni	(405)	(4.2)
Cr	−41	−0.42	Cu	(444)	(4.6)
Mn	154	1.60	Zn	(675)	(7.0)

* Estimated figures are in brackets.

It is more usual, however, to discuss the thermodynamics of aqueous redox reactions in terms of the **standard electrode potential** or **standard redox potential**, $E^\ominus$, rather than using $\Delta G_m^\ominus$ values.

2.1 The standard redox potential

Thermodynamic stabilities with respect to oxidation are conveniently compared using the standard redox potential or standard electrode potential $E^\ominus$.

Note that $E^\ominus$ values refer to equations in which the oxidised state is *always* written on the left, so we shall first reverse Reaction 2.1:

$$M^{3+}(aq) + \tfrac{1}{2}H_2(g) = M^{2+}(aq) + H^+(aq) \tag{2.2}$$

It is usual when discussing redox potentials to abbreviate Equation 2.2 to

$$M^{3+}(aq) + e = M^{2+}(aq) \tag{2.3}$$

where e is shorthand for $\frac{1}{2}H_2(g) - H^+(aq)$. The e is not strictly an electron, but it is often thought of as the number of electrons transferred in the reaction (which we write with a superscript minus, e^-). However, by thinking of it as an electron, you can see that the total charge of the combination $[\frac{1}{2}H_2(g) - H^+(aq)]$ is −1 and so confirm that Equation 2.3 is balanced like Equation 2.2.

The system in Equation 2.2 is called a **couple** and its standard redox potential is written $E^\ominus(M^{3+}|M^{2+})$ with the oxidised state to the left inside the bracket and the reduced state to the right. This matches their positions in Equation 2.3.

The general relationship between $\Delta G_m^\ominus$ and $E^\ominus$ for any redox couple is given by:

$$\Delta G_m^\ominus = -nFE^\ominus \tag{2.4}$$

Here n is the coefficient of e on the left-hand side of the couple (in other words the number of electrons transferred). It is also the decrease in oxidation state (always positive) between the oxidised state on the left of the couple and the reduced state on the right. F is a constant called the Faraday which has a value of 96.485 kJ mol^{-1} V^{-1}. Values of $E^\ominus$ are recorded in volts, V.

In Equations 2.2 and 2.3, $n = 1$, so if we write the respective $\Delta G_m^\ominus$ values as $\Delta G_m^\ominus(2.2)$ and $\Delta G_m^\ominus(2.3)$, then

$$\Delta G_m^\ominus(2.2) = \Delta G_m^\ominus(2.3) = -FE^\ominus(M^{3+}|M^{2+})$$

In the previous section, we considered the reaction

$$M^{2+}(aq) + H^+(aq) = M^{3+}(aq) + \tfrac{1}{2}H_2(g) \tag{2.1}$$

This is the reverse of Equation 2.2 and so

$$\Delta G_m^\ominus(2.1) = -\Delta G_m^\ominus(2.2) = FE^\ominus(M^{3+}|M^{2+}) \tag{2.5}$$

Both quantities $E^{\ominus}(M^{3+}|M^{2+})$ and $\Delta G_m^{\ominus}(2.1)$ for the first-row transition metals are given in Table 2.1. You should satisfy yourself that the two quantities are correctly related by Equation 2.5.

Equation 2.5 shows that $\Delta G_m^{\ominus}(2.1)$ is proportional to $E^{\ominus}(M^{3+}|M^{2+})$ and you can see that both quantities are relatively large and positive when oxidation of M^{2+} is difficult or reduction of M^{3+} is relatively easy. For example, experiments that you watched in Activity 1.4 showed that it is hard to oxidise Co^{2+}(aq).

■ How is this reflected by the values of $\Delta G_m^{\ominus}(2.1)$ and $E^{\ominus}(M^{3+}|M^{2+})$ in Table 2.1?

□ For cobalt, both are large and positive. Each of these results indicates that oxidation of Co^{2+}(aq) to Co^{3+}(aq) is difficult and that reduction of Co^{3+}(aq) to Co^{2+}(aq) is relatively easy.

A list of $E^{\ominus}$ values in acid solution is given in the Appendix. Roughly speaking, redox systems with $E^{\ominus}$ values greater than about 1.1 V in acid solution contain powerful oxidising agents; conversely, those with $E^{\ominus}$ values less (more negative) than about −0.1 V in acid solution contain powerful reducing agents. The more powerful the oxidising agent in a system, the more positive is $E^{\ominus}$.

Now look again at the values of $E^{\ominus}(M^{3+}|M^{2+})$ in Table 2.1.

■ Does the thermodynamic stability of M^{2+}(aq) with respect to M^{3+}(aq) increase across the series?

□ The negative values of $E^{\ominus}(M^{3+}|M^{2+})$ at the beginning of the series and the positive values at the end show that, overall, it gets harder to oxidise M^{2+}(aq) as we move from scandium to zinc. In fact, the thermodynamic stability of M^{2+}(aq) increases from element to element except from vanadium to chromium, where $E^{\ominus}$ falls very slightly from −0.26 V to −0.42 V, and between manganese and iron, where there is a much bigger fall from 1.60 V to 0.77 V. We detected this second exception in the experiments you watched in Activity 1.4.

Redox potentials can be used to predict whether or not a particular redox reaction is thermodynamically possible under standard conditions. Consider the two redox systems:

$$Mn^{3+}(aq) + e = Mn^{2+}(aq) \quad E^{\ominus} = 1.60\ V$$

$$Cr^{3+}(aq) + e = Cr^{2+}(aq) \quad E^{\ominus} = -0.42\ V$$

■ Would you expect Mn^{3+}(aq) to be thermodynamically capable of oxidising Cr^{2+}(aq); in other words, is the reaction

$$Mn^{3+}(aq) + Cr^{2+}(aq) = Mn^{2+}(aq) + Cr^{3+}(aq)$$

likely to be to be thermodynamically favoured?

To show just how similar the variations of $\Delta G_m^\ominus$ for Reactions 2.1 and 2.13 are, they are plotted together in Figure 2.2. The parallelism between the two variations is even closer than the parallelism of either with the I_3 variation.

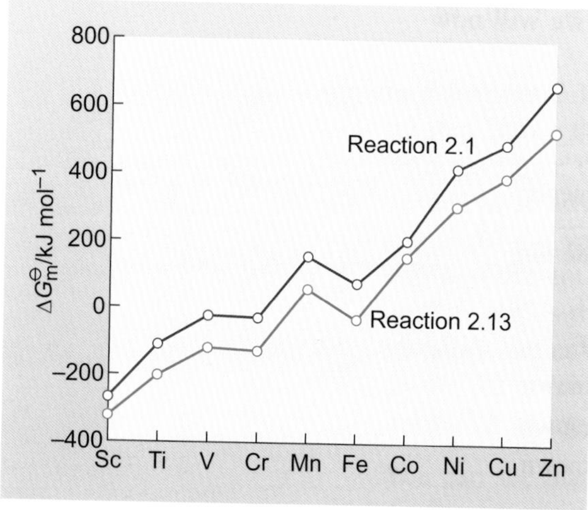

Figure 2.2 Values of $\Delta G_m^\ominus$ for Reaction 2.1 (blue) and Reaction 2.13 (green).

The increasing stability of the dichlorides can be linked, in the same way as the stability of the M^{2+}(aq) ions, with the gradually tighter binding of electrons as the nuclear charge increases in the series of M^{2+} ions with configurations [Ar]3d^n. In making this link, we are thinking in terms of dichlorides and trichlorides containing M^{2+} and M^{3+} ions, respectively.

Table 2.2 contains some revealing information on the relative stabilities of the trihalides of a particular metal. For manganese and cobalt, only the trifluoride is known at room temperature. In spite of various preparative attempts, no trichlorides, tribromides or triiodides are known for these two elements. This suggests that only the trifluorides are stable with respect to reactions such as

$$MnX_3 = MnX_2 + \tfrac{1}{2}X_2$$

$$CoX_3 = CoX_2 + \tfrac{1}{2}X_2$$

(where X is a halogen) at room temperature. Fluorine is the most effective at stabilising the higher halide with respect to the lower. Iodine is the least effective, as can be seen from the absence of FeI_3 in the column of iron trihalides. The capacity of the halogens to bring out higher oxidation states in halogen compounds decreases in the order F > Cl > Br > I.

■ What evidence is there in Table 2.2 to suggest that chlorine is better than bromine or iodine at bringing out high oxidation states?

□ $MnCl_3$ is stable below −40 °C but $MnBr_3$ and MnI_3 are unknown and we presume this is because they are even less stable with respect to the dihalides. Likewise $FeCl_3$ is known, but FeI_3 is not.

2.3 Some further comments on the relative stabilities of the dipositive and tripositive oxidation states

We have now considered five systems in which the stability of the dipositive oxidation state with respect to the tripositive oxidation state follows a similar pattern across the first transition series. The five systems are the aqueous ions, the fluorides, the chlorides, the bromides and the iodides and, in each case, the stability of the dipositive oxidation state shows a general increase across the series, coupled with a marked decrease between manganese and iron.

You will find when we are talking about the oxidation state of a transition metal species in general terms that we will represent the oxidation state in brackets as Roman numerals, as we have done here.

Such a similarity shows that when we classify simple compounds by oxidation state, we can make useful generalisations about their chemistry. To some extent, these generalisations can even be of an absolute kind. Thus, nowhere in the systems that we have examined has a compound of copper(III) or zinc(III) appeared, and we might therefore feel justified in saying that compounds of this type have low stability with respect to copper(II) or zinc (II). Such a conclusion would be correct: no compound of zinc(III) has ever been made, and it is very difficult to make compounds of copper(III) by oxidising copper(II).

However, while generalisations about the stabilities of oxidation states are useful as a first approximation, remember that the stability of a particular oxidation state can be affected by manipulating the chemical environment in which it occurs. To take a simple case, at room temperature, MnF_3 is stable with respect to MnF_2, but the unknown compound MnI_3 is almost certainly unstable with respect to MnI_2 and I_2.

Put another way, the synthesis of new or unusual oxidation states requires an ability to defeat the kind of pessimistic generalisation that we made above about zinc(III) and copper(III). Such an ability was displayed in the video clip you watched in Activity 1.3 when a compound of copper(III) was synthesised in the form of the solid fluoride compound, K_3CuF_6.

2.4 Lower oxidation states

So far in this and the previous section, we have restricted ourselves to the study of aqueous monatomic cations, aqueous oxo-ions, binary halides and oxides and solid fluoride and oxide compounds. With these restrictions, the only important oxidation state below +2 for the first transition series is +1, and at normal temperatures, copper is the only first-row transition metal to form it. We shall discuss the chemistry of copper by using redox potentials.

2.4.1 Copper

The redox potentials of an element are conveniently displayed by using a **potential diagram**. The potential diagram for copper in acid solution is shown below:

$$Cu^{2+} \xrightarrow{\;0.16\ V\;} Cu^+ \xrightarrow{\;0.52\ V\;} Cu$$
$$Cu^{2+} \xrightarrow{\qquad0.34\ V\qquad} Cu$$

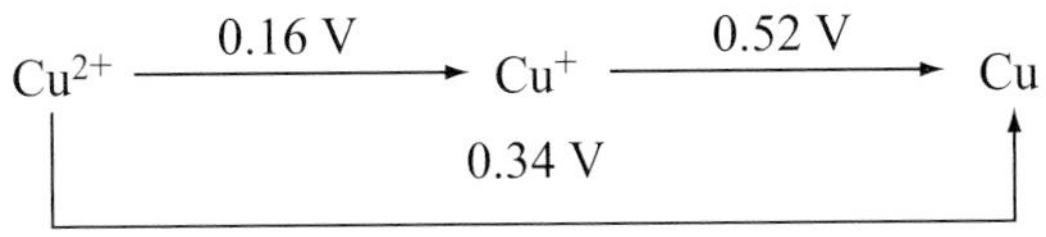

The most prominent species in the various oxidation states are written down in a chain, with the most oxidised state on the left and the most reduced on the right. The two oxidation states in any couple are connected by an arrow pointing from left to right and this is labelled with the standard electrode potential of that couple.

From the diagram you can see that the value of $E^{\ominus}(Cu^+|Cu)$ is 0.52 V and is greater than the $E^{\ominus}$ value involving $Cu^{2+}(aq)$ and $Cu^+(aq)$, $E^{\ominus}(Cu^{2+}|Cu^+)$, which is 0.16 V. Therefore, $Cu^+(aq)$ going to $Cu(s)$ oxidises $Cu^+(aq)$ to $Cu^{2+}(aq)$. In other words, $Cu^+(aq)$ is thermodynamically unstable with respect to the reaction

$$2Cu^{+}(aq) = Cu(s) + Cu^{2+}(aq) \quad (2.14)$$

This is a **disproportionation** reaction.

The result that we have just obtained is quite general. If the couple linking an ion or compound with a lower oxidation state has an $E^{\ominus}$ value greater than that of a couple linking it with an upper oxidation state, the ion or compound is thermodynamically unstable to disproportionation.

In this particular case, $Cu^{+}(aq)$ is colourless and does indeed decompose according to Equation 2.14; consequently only very dilute solutions of it can be obtained. However, copper(I) compounds can be stabilised if the equilibrium is shifted to the left by adding an anion that forms an insoluble compound with $Cu^{+}(aq)$.

CuI is just such a compound. If aqueous iodide ions are added to an aqueous solution of copper(II) sulfate, a white precipitate of CuI is formed and a deep brown colour appears in the solution. This is iodine, which means that the iodide has been oxidised.

$$Cu^{2+}(aq) + 2I^{-}(aq) = CuI + \tfrac{1}{2}I_2(s)$$

If excess iodide is present, the iodine dissolves to form the ion $I_3^{-}(aq)$.

The presence of $I^{-}(aq)$ precipitates CuI, which is sufficiently insoluble not to disproportionate according to Equation 2.14. We say that CuI(s) is *stable with respect to disproportionation* in aqueous solution.

Many simple compounds of copper(I) show a marked tendency to revert to copper(II). For example, the red copper oxide, Cu_2O, can be made by boiling an alkaline suspension of $Cu(OH)_2$ with the reducing agent hydrazine, N_2H_4, but, on standing in the atmosphere, Cu_2O slowly forms the black oxide, CuO.

In summary, among the first-row transition elements, copper is the only element to form simple compounds such as halides, salts or oxides in the +1 oxidation state. The behaviour of these compounds suggests that if similar compounds could be made for other first-row transition metals, they would quickly revert to higher oxidation states, either by atmospheric oxidation or by disproportionation.

■ $E^{\ominus}(Cu^{2+}|Cu)$ is 0.34 V and $E^{\ominus}(Cu^{+}|Cu)$ is 0.52 V. Does copper react with dilute acids to give hydrogen?

□ No. $E^{\ominus}(Cu^{2+}|Cu)$ and $E^{\ominus}(Cu^{+}|Cu)$ are both positive, so the reactions

$$Cu(s) + H^{+}(aq) = Cu^{+}(aq) + \tfrac{1}{2}H_2(g)$$

$$Cu(s) + 2H^{+}(aq) = Cu^{2+}(aq) + H_2(g)$$

are both thermodynamically unfavourable. (See Section 2.1.1.)

2.5 Higher oxidation states

In Section 1.5 you encountered examples of transition metal ions in oxidation states greater than +3.

■ Looking back to Figure 1.6, which metals form stable oxidations states above +3?

□ Ti, V, Cr and Mn form stable oxidation states of +4 or above.

In this section, we will relate the stability of the higher oxidation states to the redox potentials of these elements.

2.5.1 Chromium

A potential diagram for chromium in acid solution is shown below:

$$Cr_2O_7^{2-} \xrightarrow{1.36\ V} Cr^{3+} \xrightarrow{-0.42\ V} Cr^{2+} \xrightarrow{-0.90\ V} Cr$$

In aqueous solution, the important higher oxidation state of chromium is +6; in acid, it occurs as the orange dichromate(VI) ion, $Cr_2O_7^{2-}$, unless the solution is very dilute when the ion present is hydrogen chromate(VI), $HCrO_4^-(aq)$.

The dichromate ion is a strong oxidising agent:

$$Cr_2O_7^{2-}(aq) + 14H^+(aq) + 6e = 2Cr^{3+}(aq) + 7H_2O(l) \quad E^\ominus = 1.36\ V$$

For example, it oxidises $Fe^{2+}(aq)$ to $Fe^{3+}(aq)$:

$$Fe^{3+}(aq) + e = Fe^{2+}(aq) \quad E^\ominus = 0.77\ V$$

and acidified hydrogen peroxide to oxygen:

$$O_2(g) + 2H^+(aq) + 2e = H_2O_2(aq) \quad E^\ominus = 0.69\ V$$

In the latter case, the overall reaction is:

$$Cr_2O_7^{2-}(aq) + 8H^+(aq) + 3H_2O_2(aq) = 2Cr^{3+}(aq) + 3O_2(g) + 7H_2O(l) \quad (2.15)$$

You saw this reaction during Activity 1.3. Because dichromate is such a strong oxidising agent, only the very strongest oxidising agents, such as persulfate, will convert chromium(III) into chromium(VI) in acid solution.

Another reaction that you watched in Activity 1.3 was the conversion of orange dichromate(VI) into yellow chromate(VI) by aqueous hydroxide ions:

$$Cr_2O_7^{2-}(aq) + 2OH^-(aq) = 2CrO_4^{2-}(aq) + H_2O(l)$$

This reaction is reversed on adding acid.

Thus in alkaline solution, the preferred form of chromium(VI) is $CrO_4^{2-}(aq)$. Now look again at Reaction 2.15. Let us consider it as an equilibrium system between chromium(VI) and chromium(III). In acid solution we have seen that equilibrium lies well over to the right: dichromate oxidises peroxide and is reduced to Cr^{3+}(aq). But if the solution is made alkaline by adding hydroxide ions, the equilibrium will shift. Firstly, dichromate will be converted into chromate which in alkali is the more stable form of chromium(VI), but more importantly, the hydroxide ions will react with the hydrogen ions that appear on the left-hand side of the equilibrium. Both these effects will shift the equilibrium to the left and stabilise chromium(VI).

This suggests that although oxidation of chromium(III) to chromium(VI) is difficult in acid, in alkali it may be much easier. You saw this demonstrated in Activity 1.3. When sodium hydroxide is added to Cr^{3+}(aq), a green precipitate of chromium(III) hydroxide, $Cr(OH)_3$, is formed. If this is heated with aqueous hydrogen peroxide, a yellow solution of CrO_4^{2-}(aq) is obtained:

$$2Cr(OH)_3(s) + 3H_2O_2(aq) + 4OH^-(aq) = 2CrO_4^{2-}(aq) + 8H_2O(l)$$

Thus in acid solution, chromium(VI) is reduced by hydrogen peroxide to chromium(III), but in alkaline solution, hydrogen peroxide oxidises chromium(III) to chromium(VI). The high oxidation state is stabilised in alkali.

High oxidation states in aqueous solution usually occur as oxoanions or oxocations. In reduction reactions, they lose oxygen atoms and these are taken up by hydrogen ions, which appear on the same side of the equation. H^+(aq) thus appears with a large coefficient on the same side of the equation as the high oxidation state. This is one reason why the generalisation that high oxidation states are stabilised in alkaline media is often true.

2.5.2   Manganese

A potential diagram for manganese in acid solution is shown below:

$$
\begin{array}{c}
\overset{\textstyle 1.51\ \text{V}}{\overbrace{\hspace{8cm}}}\\
MnO_4^- \xrightarrow{0.56\ \text{V}} MnO_4^{2-} \xrightarrow{2.27\ \text{V}} MnO_2 \xrightarrow{0.86\ \text{V}} Mn^{3+} \xrightarrow{1.60\ \text{V}} Mn^{2+} \xrightarrow{-1.18\ \text{V}} Mn
\end{array}
$$

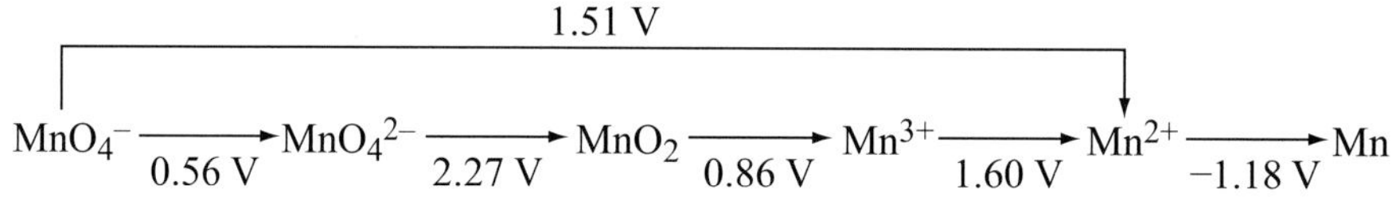

The highest oxidation state of +7 occurs as the manganate(VII) anion, MnO_4^-, commonly known as the permanganate ion. In acid solution this is a powerful oxidising agent and is usually reduced to Mn^{2+}(aq):

$$MnO_4^{2-}(aq) + 8H^+(aq) + 5e = Mn^{2+}(aq) + 4H_2O(l) \quad E^{\ominus} = 1.51\ \text{V}$$

Consequently, only the very strongest oxidising agents will convert manganese(II) into manganate(VII). In one video clip in Activity 1.3, we used sodium bismuthate, $NaBiO_3$, as an oxidising agent; however, persulfate can also be used:

$$S_2O_8^{2-}(aq) + 2e = 2SO_4^{2-}(aq) \quad E^{\ominus} = 2.00\ \text{V}$$

Persulfate only oxidises $Mn^{2+}(aq)$ to a rose-coloured solution of $Mn^{3+}(aq)$ but, if the solution is allowed to stand overnight, further oxidation occurs to purple $MnO_4^-(aq)$.

The persulfate oxidation that produced $Mn^{3+}(aq)$ was carried out with a very small initial concentration of manganese(II). If the initial concentration is increased, there is a corresponding increase in the concentration of the first product, $Mn^{3+}(aq)$, and at all but very low concentrations, $Mn^{3+}(aq)$ is unstable:

$$2Mn^{3+}(aq) + 2H_2O(l) = Mn^{2+}(aq) + MnO_2(s) + 4H^+(aq)$$

■ What type of reaction is this?

□ This reaction is another example of a disproportionation reaction. The potential diagram reveals this instability of $Mn^{3+}(aq)$ because $E^\ominus(Mn^{3+}|Mn^{2+}) = 1.60$ V is greater than $E^\ominus(MnO_2|Mn^{3+}) = 0.86$ V.

The oxidation of the insoluble MnO_2 to $MnO_4^-(aq)$ by persulfate is too slow to be observed. If a high initial concentration of $Mn^{2+}(aq)$ is used, $Mn^{3+}(aq)$ is formed initially but then disproportionates; the manganese(II) product rejoins unoxidised manganese(II) to await re-oxidation to manganese(III) and the end-product of the overall reaction is MnO_2:

$$Mn^{2+}(aq) + S_2O_8^{2-}(aq) + 2H_2O(l) = MnO_2(s) + 2SO_4^{2-}(aq) + 4H^+(aq)$$

Manganese dioxide is very often produced by the oxidation of lower oxidation states or the reduction of higher oxidation states of manganese in all solutions except the most strongly acid. In the form of pyrolusite, it is the principal ore from which manganese is extracted.

Oxidation states +4 and +7 can thus be obtained in acid solution. Oxidation states +5 and +6 must be made under alkaline conditions.

In Activity 1.3 you saw how the green manganate(VI) ion, MnO_4^{2-}, can be obtained by fusing MnO_2 with an oxygen-carrying oxidiser such as potassium nitrate:

$$MnO_2(s) + 2KOH(s) + KNO_3(s) = K_2MnO_4(s) + KNO_2(s) + H_2O(g)$$

On cooling, a green mass is obtained. It dissolves in water to give a dark green alkaline solution, which, if evaporated in a vacuum, gives dark green crystals of potassium manganate(VI), K_2MnO_4.

The dark green solution is stable only in alkali. On acidification, the manganate(VI) ion decomposes. Look at the potential diagram for acid solution shown above.

■ What disproportionation reaction might $MnO_4^{2-}(aq)$ undergo in acid solution?

- □ As $E^{\ominus}(MnO_4^{2-}|MnO_2) = 2.27\ V$ and this is greater than $E^{\ominus}(MnO_4^{-}|MnO_4^{2-}) = 0.56\ V$, MnO_4^{2-}(aq) is thermodynamically unstable with respect to disproportionation into MnO_4^-(aq) and MnO_2(s).

On acidification, MnO_4^{2-}(aq) disproportionates to give a brown suspension of MnO_2 in a purple solution of manganate(VII).

$$3MnO_4^{2-}(aq) + 4H^+(aq) = 2MnO_4^-(aq) + MnO_2(s) + 2H_2O(l)$$

Permanganate can be reduced in alkaline solution to the rather rare oxidation state, manganese(V). Addition of powdered sodium sulfite, Na_2SO_3, to a strongly alkaline solution of manganate(VII) gives a blue solution of the MnO_4^{3-} ion, which deposits blue crystals of $Na_3MnO_4.7H_2O$, sodium manganate(V) heptahydrate.

The structure of some of the higher oxidation states is considered in Box 2.1.

Box 2.1 Structural chemistry of the higher oxidation states of chromium, manganese and iron

The occurrence of MO_4^{2-} ions for the consecutive elements chromium, manganese and also iron is another example of marked similarities within a row of transition elements. They are also structurally similar.

X-ray diffraction studies of the compounds K_2CrO_4, K_2MnO_4 and K_2FeO_4, the latter containing the ion FeO_4^{2-} with iron in the (VI) state, show that all three compounds contain discrete MO_4 groupings, which can be regarded as MO_4^{2-} ions. All three MO_4^{2-} ions are tetrahedral as shown in Figure 2.3a.

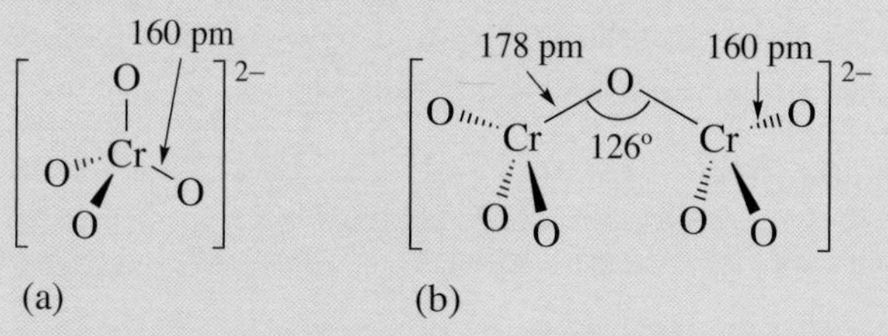

Figure 2.3 (a) Geometry of chromate(VI) ion; (b) geometry of dichromate(VII) ion.

The $Cr_2O_7^{2-}$ ion is formed by acidifying solutions of CrO_4^{2-} and X-ray diffraction studies of $(NH_4)_2Cr_2O_7$ show that two CrO_4 tetrahedra link up as shown in Figure 2.3b. The central Cr–O bonds are longer than the Cr–O terminal bonds. This would be expected because the central bridging bonds are single but the terminal bonds should have a higher order.

2.5.3 Titanium and vanadium

For chromium, manganese and iron, the oxidation states greater than +3 must be prepared, particularly in acid solution, by reacting lower oxidation states with fairly strong oxidising agents such as persulfate. In contrast, the elements titanium and vanadium have maximum oxidation states of +4 and +5, respectively, which are reasonably stable with respect to reduction; in aqueous solution, fairly powerful reducing agents are needed to convert the highest oxidation states into lower ones.

Common starting materials for experiments with titanium are the dioxide, TiO_2 (the standard pigment used in white paint), and the covalent tetrachloride, $TiCl_4$ (a colourless liquid that fumes in moist air and has been used to produce smokescreens):

$$TiCl_4(l) + 2H_2O(l) = TiO_2(s) + 4HCl(aq)$$

This reaction goes to completion when the compound is hydrolysed by a large excess of water to hydrated forms of TiO_2. The dioxide dissolves in hot concentrated sulfuric acid to give a colourless solution of titanium(IV). This is reduced by zinc metal to violet $Ti^{3+}(aq)$, which on standing undergoes atmospheric oxidation back to titanium(IV).

Titanium(II) compounds are particularly powerful reducing agents. $TiCl_2$, $TiBr_2$ and TiI_2 can be made by reducing the tetrahalides with titanium metal in non-aqueous solutions:

$$Ti(s) + TiX_4(l, s) = 2TiX_2(s)$$

TiF_2 is unknown.

Vanadium is commonly found in laboratories as the vanadium(V) compound $NaVO_3$. This dissolves in acid to give a yellow solution of vanadium(V) as $VO_2^+(aq)$:

$$NaVO_3(s) + 2H^+(aq) = Na^+(aq) + VO_2^+(aq) + H_2O(l)$$

The yellow $VO_2^+(aq)$ can be reduced further by shaking it with zinc amalgam as you saw in Activity 1.3.

2.6 Summary of oxidation-state patterns in the first transition series

Our survey of the first-row transition elements is now complete. It relied upon a classification by oxidation state. Here, therefore, you should look back at Figure 1.6, the oxidation-state chart that you developed for the elements potassium to zinc during Activity 1.3, before doing Activity 2.1 which provides a summary of this chapter and looks ahead to Chapter 3. This short activity also refers to the origin of colour in transition metal compounds, which will be considered further in Chapters 5 and 6.

Activity 2.1

You should now work through the final section of the activity *Chemistry of the transition elements*:

Section 6: Conclusion

Don't forget that there are questions on the companion website which you can use to test your understanding of the material covered in this chapter.

3 Coordination chemistry

Many of the fundamentals of coordination chemistry were discovered through the original work on cobalt complexes to which you were introduced in Activity 2.1.

If $CoCl_2.6H_2O$ and ammonium chloride are dissolved in fairly concentrated ammonia, the solution is pink. If air is bubbled through, the colour changes to deep red and, on adding concentrated HCl, a purple solid with the composition $Co(NH_3)_5Cl_3$ is precipitated. When the experiment is repeated in the presence of a charcoal catalyst, the aeration yields a yellow-brown solution, and HCl now precipitates yellow crystals of composition $Co(NH_3)_6Cl_3$: the formulae of the two precipitated compounds differ in that the first contains one less NH_3 group than the second. It is also possible to make a compound in which the number of NH_3 groups is only four: $Co(NH_3)_4Cl_3$.

When silver nitrate is added to aqueous solutions of these three cobalt compounds, different amounts of chloride are precipitated as AgCl. Every mole of $Co(NH_3)_6Cl_3$ in solution yields 3 moles of AgCl, or 100% of the dissolved chloride; with $Co(NH_3)_5Cl_3$ the yield is 2 moles or 67%, and with $Co(NH_3)_4Cl_3$, it is only 1 mole or 33%. So, how might we account for these observations?

At least two theories were proposed during 1880–1900 to explain those observations, but the one that has lasted was put forward in 1894 by Alfred Werner, a young Swiss chemist working in Zurich.

3.1 Werner's theory

Werner argued that there were *two distinct* levels of bonding within the compounds. Firstly, there is a fixed number of bonds emanating from the cobalt centre, which we now call the **coordination number**. For the cobalt–ammonia compounds this number is six. In $Co(NH_3)_6Cl_3$, the cobalt is directly linked to six NH_3 groups in a **complex unit**, which is written $[Co(NH_3)_6]$. Secondly, this complex unit is bound to three exterior chlorines to give a compound $[Co(NH_3)_6]Cl_3$. When the compound is dissolved in water, the three exterior chlorines become aqueous chloride ions, which can be precipitated by silver nitrate. This suggests that the parent compound might be ionic with the formula $[Co(NH_3)_6]^{3+}(Cl^-)_3$, and that it dissociates in water:

$$[Co(NH_3)_6]Cl_3(s) = [Co(NH_3)_6]^{3+}(aq) + 3Cl^-(aq) \quad (3.1)$$

Care should be taken not to confuse the meaning of [X] when X is a complex ion such as $[Co(NH_3)_6]$, with the other use of square brackets where they are used to represent the concentration of X. The context in which they are used should make this clear.

A unit such as $[Co(NH_3)_6]^{3+}$ is known as a complex or a **complex ion**. In $[Co(NH_3)_6]Cl_3$, the ammonias are said to occupy the **inner sphere** of the complex; the chlorines are in the **outer sphere**.

Turning now to $Co(NH_3)_5Cl_3$, one of the ammonia groups disappears and, if this theory is correct, it has been lost from the inner sphere of the complex $[Co(NH_3)_6]^{3+}$. Werner then argued that, as *the coordination number of six must be maintained*, one of the exterior chlorides must move into the inner sphere. After these changes, the ionic formulation is $[Co(NH_3)_5Cl]^{2+}(Cl^-)_2$, and the complex ion is now $[Co(NH_3)_5Cl]^{2+}$ balanced by two chloride ions. So each mole of $Co(NH_3)_5Cl_3$ dissolves to give two moles of $Cl^-(aq)$ as precipitable chloride:

$$[Co(NH_3)_5Cl]Cl_2(s) = [Co(NH_3)_5Cl]^{2+}(aq) + 2Cl^-(aq) \quad (3.2)$$

■ How did Werner write the compound $Co(NH_3)_4Cl_3$? Does the percentage of precipitable chloride implied by your answer agree with the facts?

□ To obtain a coordination number of six, two of the three chlorines must occupy inner-sphere positions. This gives $[Co(NH_3)_4Cl_2]Cl$ and the dissociation reaction.

$$[Co(NH_3)_4Cl_2]Cl(s) = [Co(NH_3)_4Cl_2]^+(aq) + Cl^-(aq) \quad (3.3)$$

This agrees with the observed fact that one-third of the chloride is precipitable.

3.1.1 Conductivity measurements of complexes

To test his theory further, Werner measured the electrical conductivities of solutions of his complexes. These conductivities usually increase with the number of aqueous ions that the complexes yield in solution. Thus, if Equations 3.1, 3.2 and 3.3 are correct, $Co(NH_3)_6Cl_3$, $Co(NH_3)_5Cl_3$ and $Co(NH_3)_4Cl_3$ yield four, three and two aqueous ions, respectively. This is corroborated by the electrical conductivities: the value for $Co(NH_3)_6Cl_3$ is the largest of the three, and that for $Co(NH_3)_4Cl_3$ is the least.

Werner made such conductivity measurements, not just for cobalt complexes containing chloride, but also for those in which chloride had been partly or wholly replaced by other halides, or by the nitrite group, NO_2^-. Some results are listed in Table 3.1. They show how the conductivities separate the compounds into well-defined groups that yield 4, 3, 2 or zero ions in solution.

Of particular importance were the results for $Co(NH_3)_6(NO_2)_3$, $Co(NH_3)_5(NO_2)_3$ and $Co(NH_3)_3(NO_2)_3$. The conductivities suggest that the first two compounds yield four and three ions in solution, respectively.

■ What does Werner's theory predict for $Co(NH_3)_3(NO_2)_3$?

□ The six-coordination of cobalt demands $[Co(NH_3)_3(NO_2)_3]$; there are no ionisable groups, so the compound should be a *molecular covalent* complex and therefore non-conducting.

This striking prediction meant that the data for $Co(NH_3)_3(NO_2)_3$ were of special importance, and the last result in Table 3.1 shows that the conductivity measurements corroborated Werner's predictions.

Table 3.1 Conductivity values of some cobalt compounds in aqueous solution.

Formula	Molar conductivity*/ $cm^2\ ohm^{-1}\ mol^{-1}$	Proposed number of aqueous ions
$[Co(NH_3)_6]Cl_3$	431.5	4
$[Co(NH_3)_6](NO_2)_3$	421.9	4
$[Co(NH_3)_5Cl]Cl_2$	261.3	3
$[Co(NH_3)_5(NO_2)]Cl_2$	246.4	3
$[Co(NH_3)_5(NO_2)](NO_2)_2$	234.4	3
$K[Co(NH_3)_2(NO_2)_4]$	99.3	2
$[Co(NH_3)_4(NO_2)_2]Cl$	98.3	2
$[Co(NH_3)_3(NO_2)_3]$	8.0†	0

* At a concentration of 0.001 mol dm^{-3}.

† The expected value for a non-electrolyte is zero but, in practice, there is a residual conductivity caused by impurities or a reaction with the solvent.

The adaptability of Werner's theory is illustrated by attempts to obtain from aqueous solutions a cobalt complex with one less NH_3 group than $[Co(NH_3)_3(NO_2)_3]$. Elimination of the NH_3 group is balanced by the appearance of a metal ion, as in $KCo(NH_3)_2(NO_2)_4$. As Table 3.1 shows, the compound is a 1:1 electrolyte. Six-coordinate cobalt accounts for all the NH_3 and NO_2^- groups; writing potassium as K^+ gives $K^+[Co(NH_3)_2(NO_2)_4]^-$, which is a 1:1 electrolyte with a complex anion $[Co(NH_3)_2(NO_2)_4]^-$.

Werner's own famous summary of his experiments is the plot of solution conductivities against the formulae of selected compounds shown in Figure 3.1. It shows with great clarity the effect of the step-by-step displacement of neutral NH_3 groups by external anionic groups entering the inner sphere of the cobalt complex.

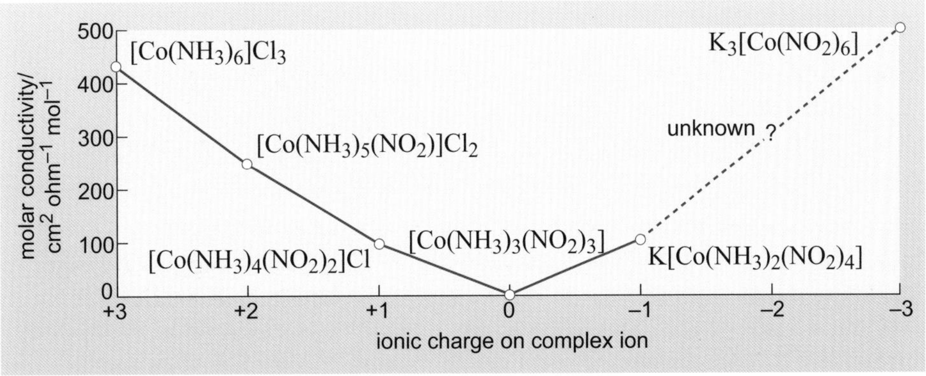

Figure 3.1 The molar conductivities of a series of cobalt complexes according to Alfred Werner and others.

- Chromium forms a series of six-coordinate complexes similar to those of cobalt. The compound $Cr(NH_3)_4(H_2O)Cl_3$ dissolves in water, and in aqueous solution has a molar conductivity of 261 $cm^2\ ohm^{-1}\ mol^{-1}$. Two-thirds of the total chlorine content can be precipitated from aqueous solution with silver nitrate. Use Werner's theory to write a formula for the compound, and an equation for its dissolution in aqueous solution.

□ $[Cr(NH_3)_4(H_2O)Cl]Cl_2$; the dissolution is:

$$[Cr(NH_3)_4(H_2O)Cl]Cl_2(s) = [Cr(NH_3)_4(H_2O)Cl]^{2+}(aq) + 2Cl^-(aq)$$

A coordination number of six is achieved by assuming that H_2O, like NH_3, can occupy an inner-sphere position. The dissolution then gives three ions, two of which are chlorides and this is consistent with the molar conductivity (Table 3.1), and with the precipitation of two-thirds of the chlorine.

3.1.2 Concluding remarks

You met the aqueous transition metal dipositive ions in Section 1.4. Although we have written these ions as $M^{2+}(aq)$, they are in fact complex ions.

■ What is the ligand in this case? Write out the formula of $Cr^{2+}(aq)$, using the inner-sphere ligands explicitly.

□ The ligand is H_2O. We write the formula of this **aquo** complex ion as $[Cr(H_2O)_6]^{2+}(aq)$.

Finally, notice that the cobalt complexes discussed so far have something more than just their octahedral coordination in common: in each case the cobalt has the same oxidation state of +3. Thus in $K[Co(NH_3)_2(NO_2)_4]$, the closed-shell configurations of the ligands are NH_3 and NO_2^-. Removing $2NH_3$ and $4NO_2^-$ from the anion $[Co(NH_3)_2(NO_2)_4]^-$ leaves Co^{3+}: the anion is a cobalt(III) complex where (III) denotes the nominal oxidation state of the metal ion if the ligands were removed.

3.2 Ligands and their nomenclature

One mark of a truly great scientific theory is its fertility: it creates a whole new research programme, generating ideas and experiments that otherwise would not happen. Werner's theory did this, principally because he made bold speculations about the geometry of his complexes. For six-coordinate cobalt, he argued that the surrounding groups were arranged at the corners of an octahedron. This arrangement is shown for the complex $[Co(NH_3)_6]^{3+}$ in structure **3.1**. A typical complex such as this consists of a central metal atom surrounded by coordinating groups called **ligands**. Table 3.2 shows some common ligands that you will frequently meet. They are listed in states with *closed-shell configurations*. Thus, with the exception of the sulfur in SO_4^{2-}, all the elements in the ligands have noble gas configurations: for example, halogens are listed as halide anions.

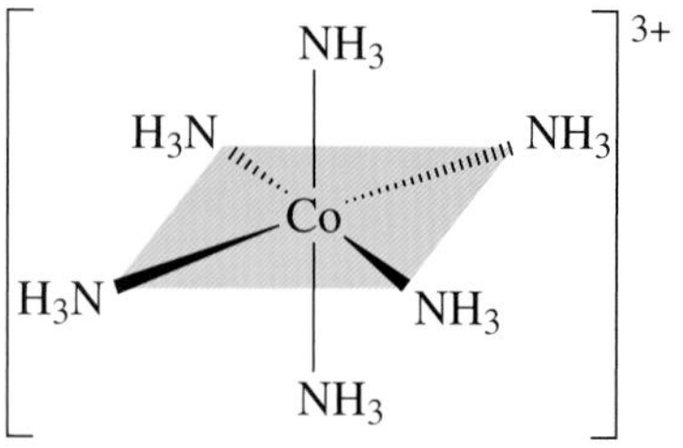

3.1

Table 3.2 Some simple ligands.

Coordinating atom	Neutral species	Ions
C	CO	CN^-
N	NH_3	NO_2^-*
O	H_2O	O^{2-}, OH^-, SO_4^{2-}, NO_2^-*
halogen		F^-, Cl^-, Br^-, I^-

* NO_2^- can coordinate through nitrogen or oxygen.

So how do ligands bind to the metal atom in a complex? In discussing the bonding, it is useful to begin with two very simple ideas.

1 There is an electrostatic contribution: all of the ligands in Table 3.2 contain atoms with **non-bonded** or **lone pairs** of electrons. The central metal atom carries a net positive charge, so the ligands orientate themselves with the negative charge of the lone pairs directed towards this positive site; the electrical interaction between them binds the metal atom and ligand together.

2 There is a covalent contribution: the ligands' lone pairs become **electron-pair bonds** (or **dative bonds**) if the ligands donate them to the metal. Compare this with the bond formed when the **Lewis base** (an electron-pair donor), NH_3, donates its lone pair to the **Lewis acid** (an electron-pair acceptor), BCl_3, in the compound $H_3N \rightarrow BCl_3$. In coordination chemistry the metal ion is acting as the Lewis acid, accepting electron pairs from the ligand.

This last example shows that the type of bonding we are considering need not be confined to metal complexes, and the class of compounds in which it is found is often given the general name, **coordination compounds**.

3.2.1 Ligands bonding through more than one atom

When ammonia acts as a ligand in a metal complex (structure **3.1**), it forms one bond to the metal atom. But there are ligands that contain two or more atoms with lone pairs of electrons, and if the distance between these atoms is neither too large nor too small, they may be able to form separate bonds to the *same* metal site.

3.2

One example is 1,2-diaminoethane (**3.2**), which is still commonly referred to by its old name ethylenediamine, and is conventionally abbreviated to en when describing a complex. This carries two nitrogen atoms arranged such that they can span two *adjacent* corners of the octahedral arrangement taken up by the ligands in cobalt complexes. Thus in structure **3.3**, three such molecules complete the octahedron by forming three five-membered rings with a common vertex at cobalt. The resulting complex, $[Co(en)_3]^{3+}$, is usually represented by the simplified structure **3.4**.

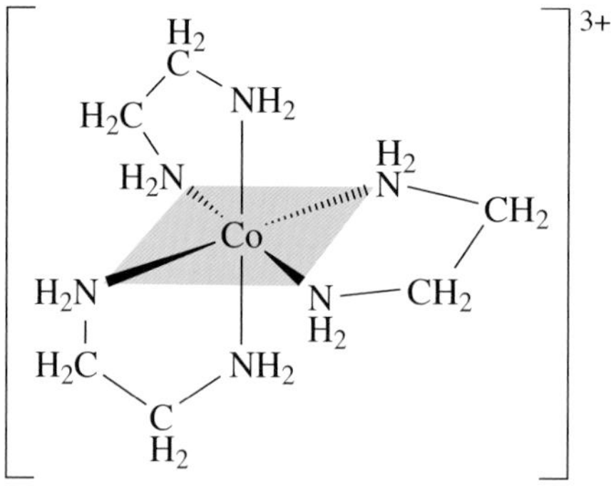

3.3

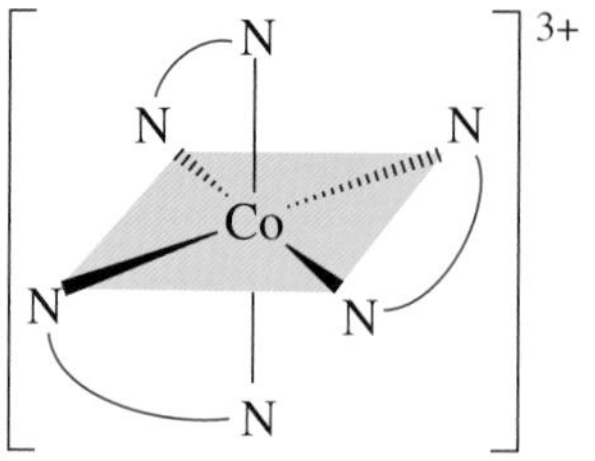

3.4

Because en has two possible points of attachment through which it can bind to a metal atom, it is known as a **bidentate ligand**. This distinguishes it from **monodentate ligands** such as NH_3, H_2O and halides that bind through just *one* atom. Table 3.3 shows some other bidentate ligands. They include 1,10-phenanthroline (phen) and the negative ion, acetylacetonate ($acac^-$). Table 3.3 also contains three **polydentate ligands** that can bind through *more than two* atoms. **Denticity** is the number of atoms in a single ligand that bind to a central metal atom. Coordinating atoms are identified using the Greek letter κ (kappa) followed by a superscript number to indicate the denticity. If the bonding groups are sufficiently far apart and of suitable configuration to form separate bonds to the same metal, as in these examples, the ligand is said to be **chelating**, and the resulting complex is called a **chelate**. Note especially the important hexadentate ligand ethylenediaminetetraacetate ($edta^{4-}$) shown as a complex with iron in structure **3.5**.

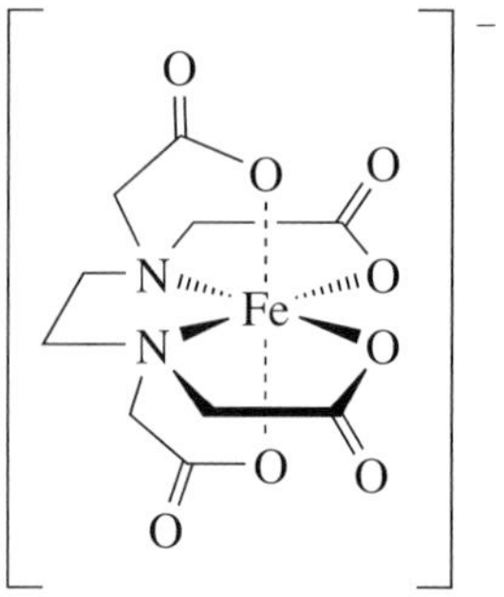

3.5

Developing this idea further, there exists a large family of **macrocyclic** ligands, which are polydentate ligands in which the bonding atoms are constrained in a planar arrangement around a central metal atom; you will meet an example (porphyrin) in Section 3.4.3.

A further term particularly applied to organic ligands, and compounds and complexes containing metal-carbon bonds, but not exclusively, is **hapticity**, which is defined as the number of contiguous atoms (connected to each other) in the ligand directly bonded to the metal. If two or more connected atoms of the ligand are directly bonded to the same metal atom the ligand is said to be *polyhapto*. The number of ligand atoms directly bonded to the metal atom is indicated by the Greek letter η (eta), followed by a superscript to indicate the hapticity. As an example, the compound ferrocene (dicyclopentadienyliron) (**3.6**), a member of a class of compounds called metallocenes, consists of two cyclopentadienyl rings bound through all five of their carbon atoms to a central iron atom; these are η^5 ligands.

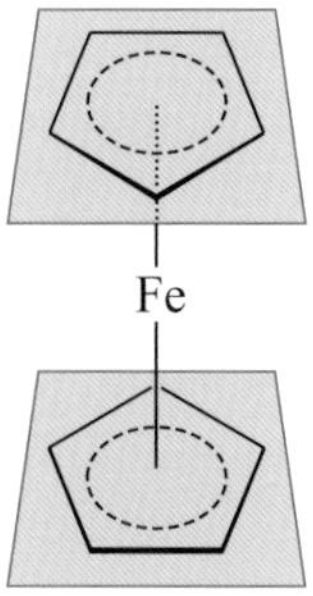

3.6

■ What is the oxidation state of the metallic element, and the associated d-electron configuration, in $[Cr(NH_3)Cl(en)_2]Cl_2$ and $[Ni(edta)]^{2-}(aq)$?

□ The oxidation states are chromium(III) and nickel(II); they are associated with the d-electron configurations $3d^3$ and $3d^8$, respectively. For $[Cr(NH_3)Cl(en)_2]Cl_2$, the closed-shell configurations of the ligands are en, NH_3 and Cl^-. Removal of 2en, NH_3 and $3Cl^-$ leaves Cr^{3+}, which has the configuration $[Ar]3d^3$. With $[Ni(edta)]^{2-}$, the closed-shell configuration of the ligand is $edta^{4-}$. Its removal leaves Ni^{2+}, with the configuration $[Ar]3d^8$.

Table 3.3 Some common polydentate ligands. They are tabulated in their closed shell configurations. (Ligating atoms are shown in colour.)

Name (abbreviation)	Formula	Structure
bidentate		
1,2-diaminoethane (en)	$H_2NCH_2CH_2NH_2$	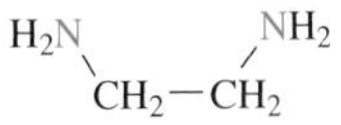
2,2′-bipyridyl (bipy)	$(C_5NH_4)_2$	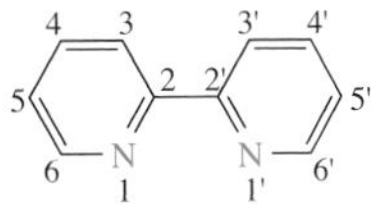
acetylacetonate ($acac^-$)	$(CH_3COCHCOCH_3)^-$	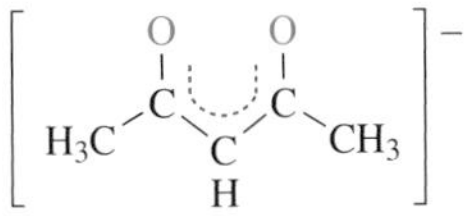
bis(diphenylphosphino)ethane (dppe)	$Ph_2PCH_2CH_2PPh_2$	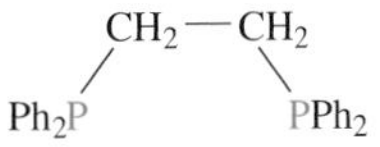
1,10-phenanthroline (phen)	$C_{12}N_2H_8$	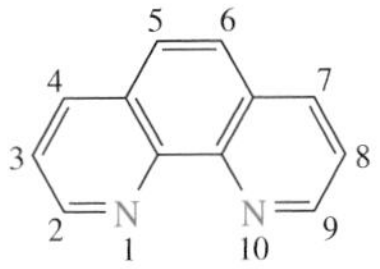
tridentate		
diethylenetriamine (dien)	$H_2N(CH_2)_2NH(CH_2)_2NH_2$	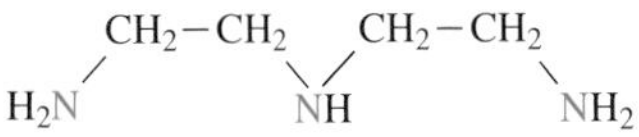
tetradentate		
triethylenetetraamine (trien)	$H_2N(CH_2)_2NH(CH_2)_2NH(CH_2)_2NH_2$	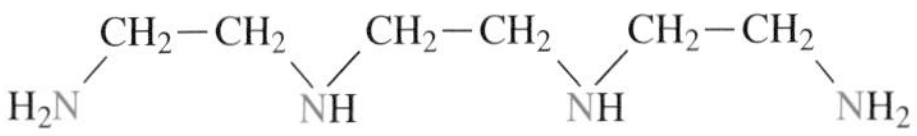
hexadentate		
ethylenediaminetetraacetate ($edta^{4-}$)	$\{(OOCCH_2)_2N(CH_2)_2N(CH_2COO)_2\}^{4-}$	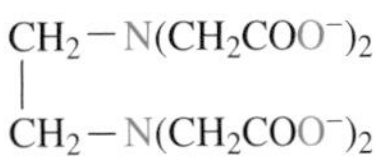

3.2.2 Multinuclear complexes and bridging ligands

Sometimes linkages can be made through a single atom to two metal centres to form a **binuclear complex**, where bridging in such a complex is identified by the Greek letter μ (mu), followed by a subscript showing how many atoms are bridged by that ligand. Such bridging linkages can be formed by ligands that have been considered to be monodentate, as shown in structures **3.7** and **3.8**. In structure **3.7**, the chloride anion is present both as a **bridging ligand** (between the two platinum atoms, i.e. μ_2) and as a **terminal ligand** (attached to one platinum atom only). Similarly, in structure **3.8**, the carbon monoxide ligand (carbonyl group) is present both as a bridging and as a terminal ligand.

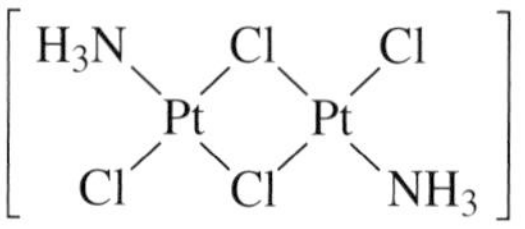

3.7

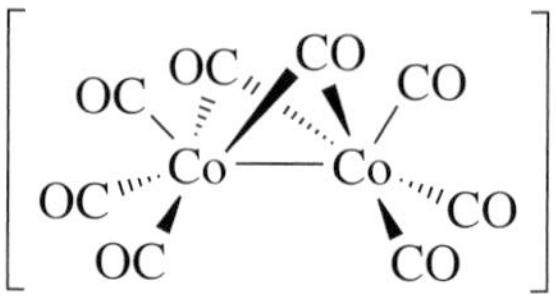

3.8

3.3 Stereochemistry and isomerism

As you work through this section and Section 3.4, you may find it useful to build molecular models to better appreciate aspects of the stereochemistry of the complexes under discussion.

Werner's theory of metal complexes included the idea that the ligands around the metal site had a particular arrangement in space. For cobalt(III) complexes, he believed that the arrangement was octahedral. To support this idea he drew an analogy with organic chemistry: the tetrahedral arrangement of bonds around carbon generates stereoisomers; would the octahedral distribution around cobalt do the same?

3.3.1 Geometric isomerism

In 1889, five years before Werner's theory was published, two compounds with the formula $CoCl_3(en)_2$ had been synthesised. One was green and the other was violet. The violet compound can be made by evaporating an aqueous solution of the green one. If it is then treated with hydrochloric acid, the green compound is regenerated. Both compounds have just one chloride that can be precipitated by $AgNO_3$.

To explain the existence of these two compounds, Werner wrote both as $[CoCl_2(en)_2]Cl$, and argued that there was a difference in the distribution of the ligands within the octahedron around cobalt (structures **3.9** and **3.10**). In the violet compound, the two chlorides occupied adjacent or *cis*-positions in the octahedron (structure **3.9**); in the green compound, the two chlorides had opposite or *trans*-positions (structure **3.10**). Thus, the two compounds were **geometric isomers**: *cis*-$[CoCl_2(en)_2]Cl$ and *trans*-$[CoCl_2(en)_2]Cl$.

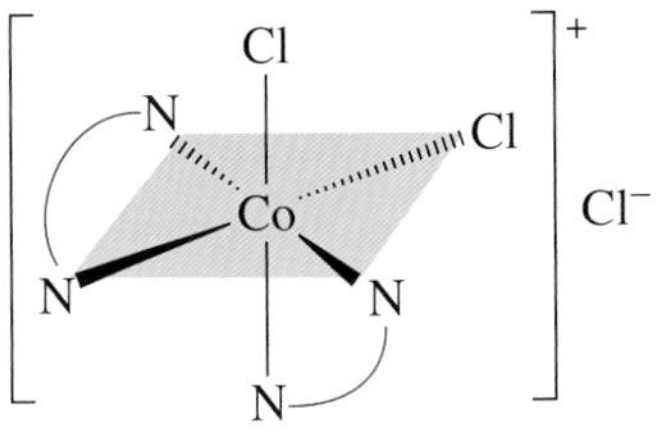

violet

3.9

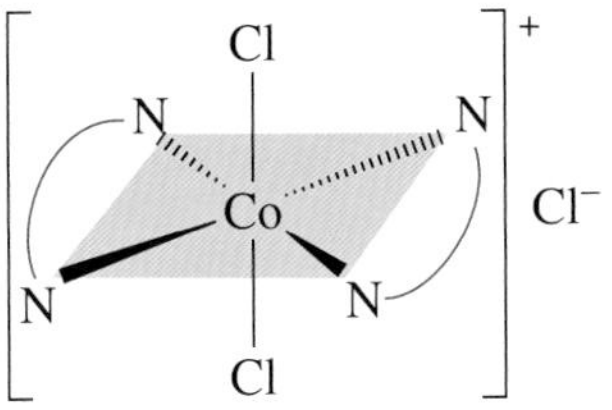

green

3.10

■ Now suppose that the two en ligands are replaced by four NH_3 molecules to give the formula $Co(NH_3)_4Cl_3$. How many compounds do Werner's ideas predict?

□ Again there are two: *cis*- and *trans*-$[Co(NH_3)_4Cl_2]Cl$ (structures **3.11** and **3.12**, respectively

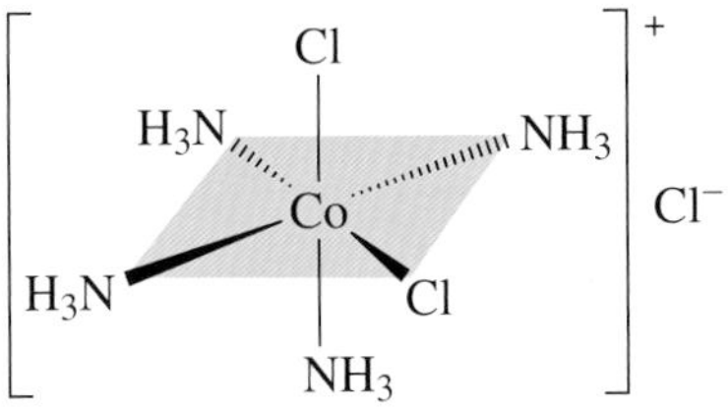

violet

3.11

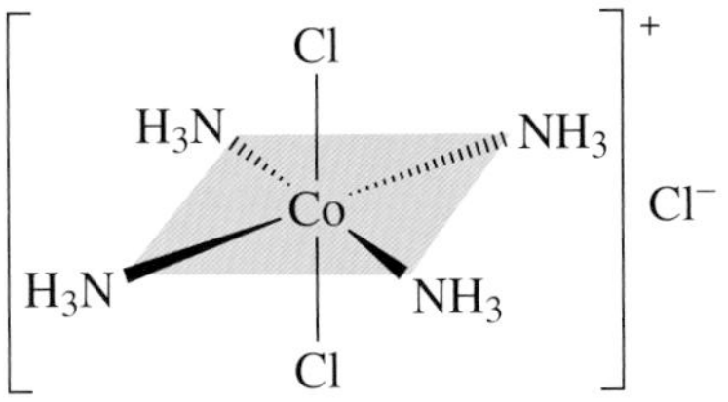

green

3.12

When Werner published his theory, just one compound of formula $[Co(NH_3)_4Cl_2]Cl$ was known, and it was green. However, in 1907, after many disappointments, Werner confirmed his prediction by preparing another compound with the same formula. This isomer was violet, and Werner assumed that it was *cis*-$[Co(NH_3)_4Cl_2]Cl$, the counterpart of the green *trans* compound that was already known. Indeed, modern X-ray crystallography has entirely confirmed the structures that Werner proposed.

Similar isomerism occurs with octahedral molecules of the type $[MA_3B_3]$, where A and B are ligands. Here, identical ligands can be arranged so that they are all on the same 'side' or 'face' of the complex – the **facial isomer** (*fac*), **3.13**, or they may span the complex so that one is *trans* to one of the other two to yield the **meridional isomer** (*mer*), **3.14**.

A convenient way of drawing these isomers is to imagine a view down an axis through a face (as you look into the page). This leads to the representations **3.15** and **3.16** for the facial and meridional isomers, respectively.

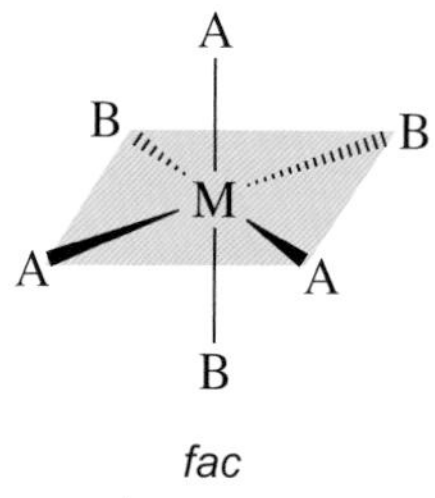

3.13

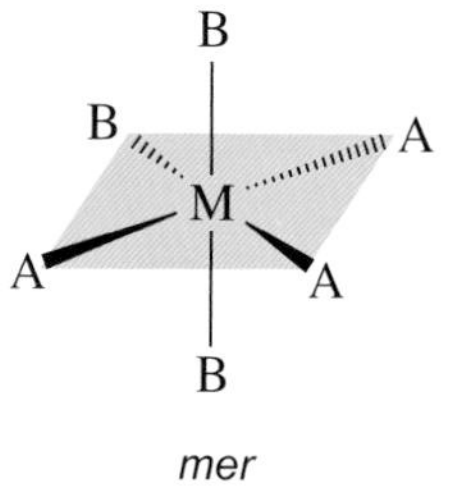

3.14

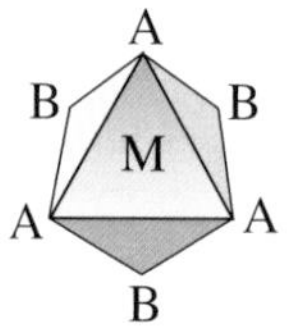

3.15

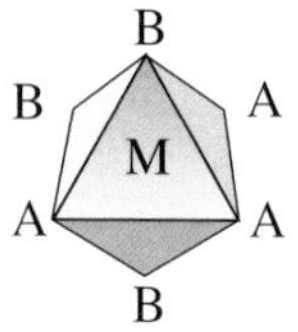

3.16

Structure **3.17** shows the complex *fac*-tris(acetonitryl)tricarbonyltungsten(0).

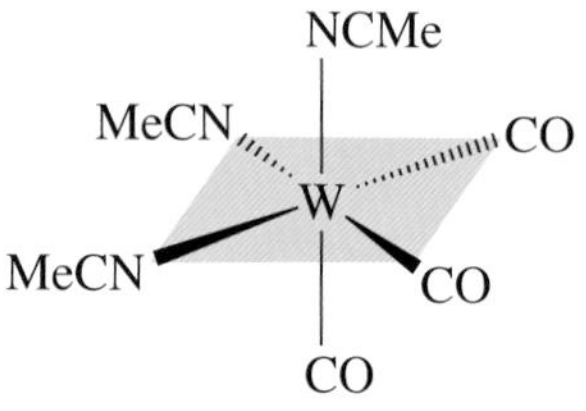

3.17

3.3.2 Optical isomerism

Another example of the fertility of Werner's theory was that it raised the possibility of **optically active transition-metal complexes**. Consider compounds with the formula [Co(NH_3)Cl(en)$_2$]Br_2. The complex cation can exist in both *cis* and *trans* forms. Look first at the *trans* form (structure **3.18**).

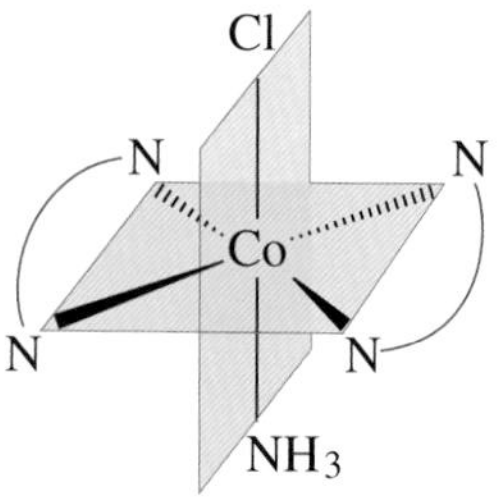

3.18

■ Does this have any planes of symmetry? (Assume that there is free rotation in the NH_3 about the N–metal bond.)

□ Yes: for example, the vertical plane shown in structure **3.18**, which contains Cl, NH_3 and Co.

The presence of a plane of symmetry means that the compound is **achiral**. But in *cis*-[Co(NH_3)Cl(en)$_2$]$^{2+}$ there is no such plane of symmetry: the complex is **chiral**, and the two non-superimposable mirror images of structures **3.19** and **3.20** correspond to two different compounds, which rotate the plane of polarised light in opposite directions. Thus, in principle, three different stereoisomers with the formula [Co(NH_3)Cl(en)$_2$]Br_2 exist – the *trans* compound and two optical isomers of the *cis* compound.

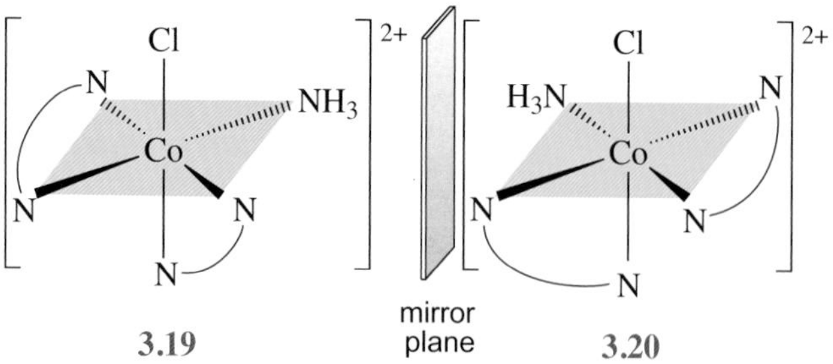

3.19 3.20

In 1911, Werner and one of his research students, Victor King, succeeded in resolving the *cis* compound into its two distinct chiral forms. This was the first time that optical isomers of a metal complex had been prepared. Werner was awarded the Nobel Prize for Chemistry in 1913.

3.3.3 Isomerism in four-coordinate complexes

So far, this section has concentrated on octahedral complexes because they were the source of important ideas about isomerism. However, those ideas are applicable to other geometries and here we shall consider four-coordination. Four-coordinate transition-metal complexes can be either square planar or tetrahedral. In some cases, the existence of isomerism allows us to distinguish between the two.

Platinum forms a series of compounds whose formulae and molar conductivities in solution are shown in Table 3.4.

Table 3.4 Conductivity values of some platinum compounds in aqueous solution at a concentration of 0.001 mol^{-1}.

Compound	**Molar conductivity/cm^2 ohm^{-1} mol^{-1}**
$Pt(NH_3)_4Cl_2$	261
$Pt(NH_3)_3Cl_2$	116
$Pt(NH_3)_2Cl_2$	2
$KPt(NH_3)Cl_3$	107
K_2PtCl_4	267

A platinum coordination number of four is common to all the formulae $[Pt(NH_3)_4]Cl_2$, $[Pt(NH_3)_3Cl]Cl$, $[Pt(NH_3)_2Cl_2]$, $K[Pt(NH_3)Cl_3]$ and $K_2[PtCl_4]$. In aqueous solution, these compounds will yield 3, 2, 0, 2 and 3 ions, respectively. For example:

$$[Pt(NH_3)_4]Cl_2(s) = [Pt(NH_3)_4]^{2+}(aq) + 2Cl^-(aq)$$

Reference to Table 3.1 shows that these numbers of ions are consistent with the molar conductivities in Table 3.4.

In the example above, we produced four-coordinate formulations for the set of platinum compounds in Table 3.4 without specifying the geometry of the complexes. The compound $[Pt(NH_3)_2Cl_2]$ exists in two forms, neither of which is optically active. This indicates that the two isomers of $[Pt(NH_3)_2Cl_2]$ are square-planar complexes in the *cis*- and *trans*-forms (**3.21** and **3.22**, respectively). If the coordination around the platinum were changed from square-planar to tetrahedral, the *cis*- and *trans*-forms would become identical.

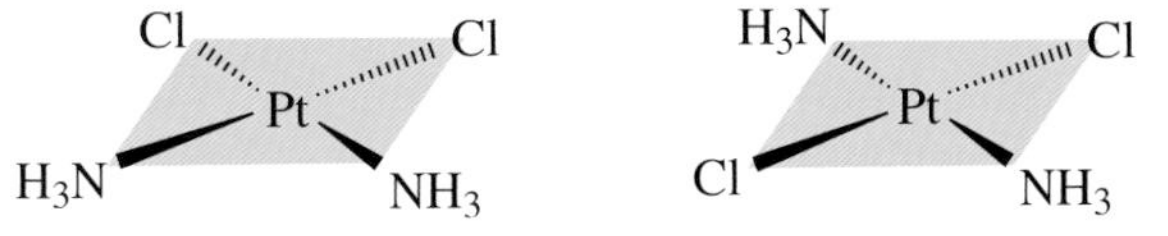

3.21 **3.22**

Notice that whether $[Pt(NH_3)_2Cl_2]$ is tetrahedral or square-planar, there are planes of symmetry and, therefore, no optical isomers. In the tetrahedral case, there is a clear connection with organic chemistry. A tetrahedral complex containing two types of ligand, $[MA_2B_2]$, cannot possess **enantiomers**; only if all four ligands are different, as in [MABCD], is this possible. A more elegant way of creating enantiomeric forms in tetrahedral coordination is to use an unsymmetrical bidentate ligand. Consider the two complexes shown in **3.23** and **3.24**. Only the bis(benzoylacetonato) complex (**3.23**) and not the bis (acetylacetonato) complex (**3.24**), which has a plane of symmetry, is optically active.

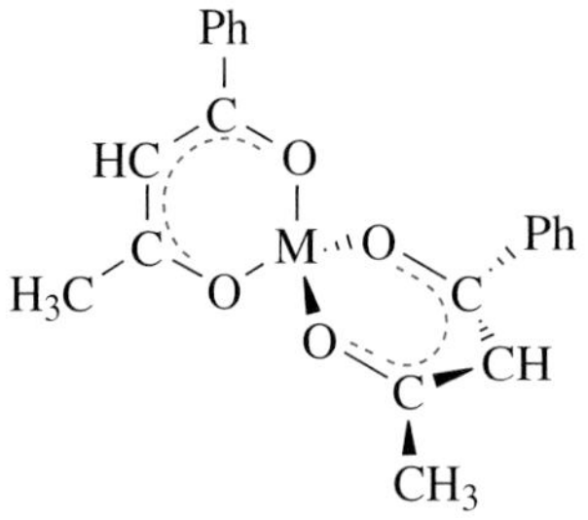

3.23

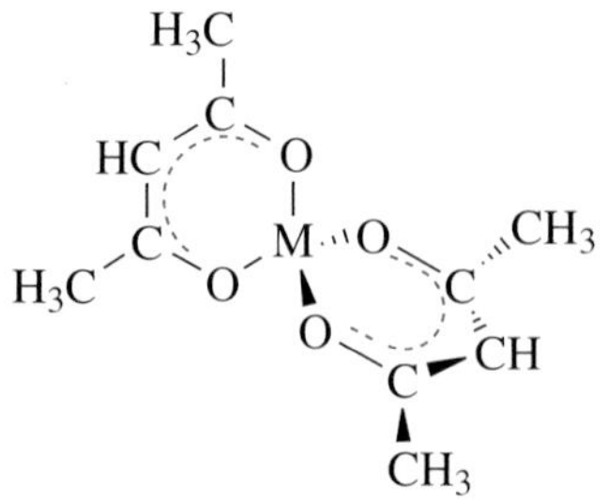

3.24

- ▪ How many isomers are possible for the octahedral complex $[MA_4B_2]$?

- □ Two – one isomer has the B groups *cis* and the other has the B groups *trans*.

3.3.4 Linkage isomerism

We conclude with a type of isomerism that is brought about by the orientation of the ligand. In organic compounds there are several substituent groups that bond to carbon through different atoms. The CN group can form nitriles (R–C≡N) and isonitriles (R–N≡C), and SCN is found both as thiocyanates (R–S–C≡N) and as isothiocyanates (R–N=C=S). There are similar possibilities in coordination chemistry where either end of the ligand, termed an **ambidentate ligand**, can bond to the metal; this is referred to as **linkage isomerism**. For example, metal complexes containing the nitrite anion, NO_2^-, can be bonded as either a nitro (**3.25**) or nitrito (**3.26**) ligand.

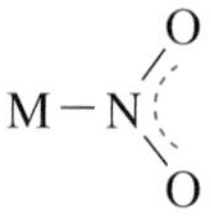

3.25

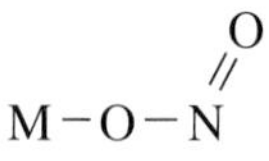

3.26

- ▪ Can you suggest how the two complexes containing these different linkages may be distinguished?

- □ Infrared spectroscopy can be used to establish the nature of the atom linked to the metal (either M–O or M–N stretches), and the N–O stretches are unique in the two compounds. The nitro isomer shows an N–O stretch at 1065 cm^{-1}, and the double-bonded nitrito at 1310 cm^{-1}.

X-ray diffraction studies of another ambidentate ligand $[SCN]^-$ have shown that for complexes of N-bonded isothiocyanate the M–N=C=S system is usually linear but, in the thiocyanate M–S–C≡N, the M–S–C angle is typically around 109°. The C–S stretching frequency is at around 820 cm^{-1} for the M–NCS system and around 700 cm^{-1} for the M–SCN system.

In the palladium complex (**3.27**), both SCN^- ligands are S-bonded even though the bent configuration is sterically more demanding. The small P–Pd–P angle of 73° results in the phenyl (Ph) groups being well clear of the area of the isothiocyanate ligands. However, if the number of carbon atoms between the two phosphorus atoms is increased from one to three, the P–Pd–P bond angle increases and the phenyl groups restrict the space available to the SCN^- ligands. (A model of the molecule will help to convince yourself of this.) The result is that the ligands tend to revert to the less sterically demanding, linear N-bonding situation (**3.28**), although the corresponding S-bonded complex can be isolated.

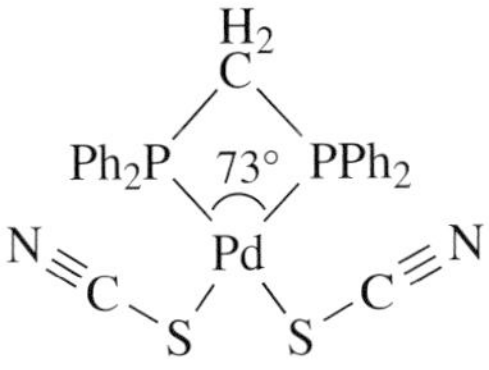

3.27

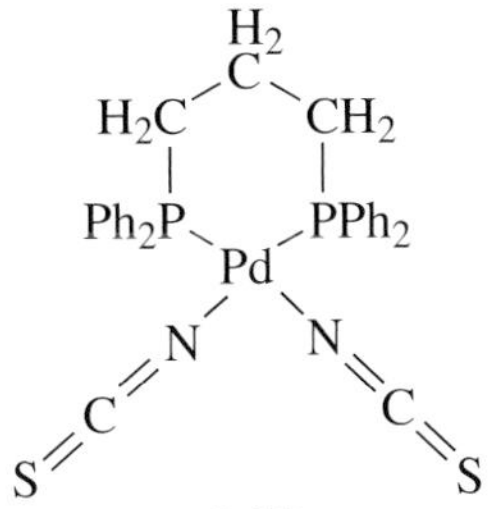

3.28

■ Using only a steric argument, which is likely to be the favoured bonding situation for the SCN^- ligand (X) in the two complex ions $[Co(NH_3)_5X]^{2+}$ and $[Co(CN)_5X]^{3-}$?

□ The cyanide ion (CN^-) or cyano ligand forms a linear M–C–N system and is thus much less sterically demanding than the ammino (NH_3) ligand, which has three hydrogen atoms attached to nitrogen. Thus, the thiocyanato ligand in the pentacyano complex is not necessarily constrained to adopt a linear structure, as it would probably have to in the pentaammine complex, but can take a bent M–S–C≡N form.

3.4 Ligands and their bonding

The transition metals form complexes with most of the non-metal atoms of the Periodic Table. So far, our attention has been directed towards ligands such as NH_3, H_2O, en, and Cl^- that bond to metals through highly electronegative atoms. In this section we shall look at the wider range of both ligands and donor atoms that is summarised in Table 3.5. What do these ligands have in common? It appears that the coordinating atom is a main-Group element from the right-hand side of the Periodic Table and that ligands are either neutral molecules or negatively charged; no cations are featured.

3.4.1 Hydrogen and elements of the early Groups

Most of the ligands that you have met so far can be regarded as two-electron donors. The hydrogen atom has just one electron and, therefore, it is the hydride ion, H^-, that acts as a two-electron ligand. There are few well-characterised complexes of transition elements with only hydride ligands; one such is the anion $[ReH_9]^{2-}$. However, there are many examples of complexes with a metal–hydrogen bond in addition to bonds between the metal and other ligands.

Table 3.5 Some typical ligands for various coordinating atoms. (You have already met some of these ligands in earlier sections, Tables 3.3 and 3.4.)

Coordinating atom	Neutral species*	Ligand name†	Ion*	Ligand name†
H			H^-	hydrido/hydride
C	CO	carbonyl	R^-	alkyl group, R
	C_2H_4	ethene	CH_3^-	methyl, Me
	C_6H_6	benzene	$CH_2CH_3^-$	ethyl, Et
			$C_6H_5^-$	phenyl, Ph
	CS_2	carbon disulfide	CN^-	cyano/cyanide
N	N_2	dinitrogen	NO_2^-	nitro or nitrito/nitrite
	NO	nitrosyl	N_3^-	azido/azide
	NH_3	ammino/ammonia	NCS^-	thiocyanato/thiocyanate
	NR_3	amino/amine	NH_2^-	amido/amide
	N_2H_4, N_2R_4	hydrazino/hydrazine	NH^{2-}	imido/imide
	$H_2NCH_2CH_2NH_2$	1,2-diaminoethane, en	N^{3-}	nitrido/nitride
P	PH_3, PR_3, PX_3	phosphino/phosphine		
	$Ph_2PCH_2CH_2PPh_2$	bis(diphenylphosphino)ethane, dppe		
As	R_3As	arsino/arsine		
Sb	R_3Sb	stibino/stibine		
O	R_2O	ether	OH^-	hydroxo/hydroxide
			O_2^{2-}	peroxo/peroxide
			CO_3^{2-}	carbonato/carbonate
	H_2O	aquo/water	$S_2O_3^{2-}$	thiosulfato/thiosulfate
			$C_2O_4^{2-}$	oxalato/oxalate, ox
			$[CH_3COCHCOCH_3]^-$	acetylacetonate, acac$^-$
S	R_2S	thioether	SCN^-	isothiocyanato/ isothiocyanide
F			F^-	fluoro/fluoride
Cl			Cl^-	chloro/chloride
Br			Br^-	bromo/bromide
I			I^-	iodo/iodide

* R represents an organic group, and X a halogen.

† Where two names are given, the name on the left is the name used in naming complexes, the name on the right is the name of the free ligand. For example, you would refer to six cyanide groups in a cobalt(II) complex, but would name the complex as hexacyanocobalt(II).

The hydride ligand occurs in at least three distinct bonding situations. The complex cation pentaamminohydridoruthenium(III), $[Ru(NH_3)_5H]^{2+}$, has ruthenium bonded directly to the hydride ligand in a terminal position. In these complexes, the length of the metal–hydrogen bond can be well approximated by the sum of the single covalent bond radii. Terminal hydride–metal bonds typically lie in the range 150–170 pm (where 1 pm = 1×10^{-12} m), as the bond shows considerable covalent character. Note that the oxidation state defined here as (III) is only used as a way of grouping transition-metal complexes by d-electron configuration. The metal is a Lewis acid accepting an electron pair from the ligand. Thus the oxidation state (III) in this example, refers to the oxidation state of the metal if the ligands were to be removed. It does not imply a Ru^{3+} cation.

The hydride ligand can act in a bridging role in complexes where the ligand is attached to two or more metal atoms. In the binuclear complex anion $[Mo_2Cl_8H]^{3-}$, the molybdenum atoms are linked by a multicentre hydrogen bond (you may have encountered the boron hydrides where an analogous situation exists). In such complexes, the M–H–M bond angles can vary through a wide range.

Bridging M–H groupings have a characteristic frequency of vibration, which falls between 1200 cm^{-1} and 1700 cm^{-1}.

■ Will terminal M–H vibrations fall at higher or lower frequency than bridging M–H?

□ Terminal M–H bonds tend to be stronger than bridging M–H bonds. The two electrons from the hydride ligand can be viewed as forming a single covalent bond in the terminal position, but these two electrons have to form a bond between hydrogen and *two* metal atoms when the hydrogen is bridging – a three-centre bond. As a result, the bond force constant is higher for terminal M–H bonds, and consequently these vibrational frequencies are seen in the higher range 1500–2100 cm^{-1}.

The third category of hydride environment is seen in cluster complexes: these are complexes where more than two metal atoms are linked to one another and to other ligands. The rhenium complex, $[Re_4(CO)_{12}H_4]$, shown in structure **3.29**, has each hydrogen bonded to three rhenium atoms.

A major difficulty arises in the structural investigation of hydride complexes. As the ligand is associated with a very low electron density, it is extremely difficult to detect directly using X-ray diffraction methods. Neutron diffraction is a much more successful technique, but this is available only in specialised national facilities. Infrared and Raman spectroscopy can be useful, but sometimes the M–H vibrations are masked by other vibrations in the complex. Fortunately, the technique of NMR spectroscopy can not only detect hydride ligands in complexes but also give detailed information on the precise environment of the ligand.

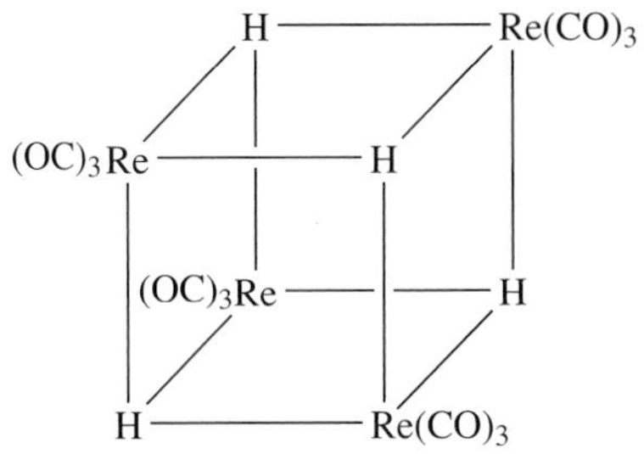

3.29

A fascinating example of the bonding of a hydride ligand occurs in the rhodium cluster, $[Rh_{13}(CO)_{24}H_3]^{2-}$. Here, NMR apparently shows one type of hydrogen atom bonded to all 13 rhodium atoms: the hydrogen atoms are rapidly moving around the polyhedral metal framework – such structures are said to be **fluxional**.

In general, the elements of the main Groups 1, 2 and 13 do not act as ligands, with the exception of a few examples involving boron.

3.4.2 Elements of Group 14

Here we only consider complexes with 'traditional inorganic' carbon ligands, such as carbonyl (CO), cyanide (CN^-) and carbon disulfide (CS_2). The ligands CO and CN^- are extremely important in transition-metal chemistry, and have the distinctive property of being able to accept electrons from the metal. We will discuss the bonding of these **π-acceptor** ligands in Chapter 6. One feature of such ligands is their ability to stabilise complexes with the metal in low oxidation states.

The carbon disulfide ligand may bond in one of the three ways shown in structures **3.30, 3.31** and **3.32**. In each example, the ligand is linked to the metal, M, through two sites. Carbon disulfide can also act as a monodentate ligand, as shown in structure **3.33**, but in these circumstances it bonds to the metal only through the sulfur atom rather than through carbon.

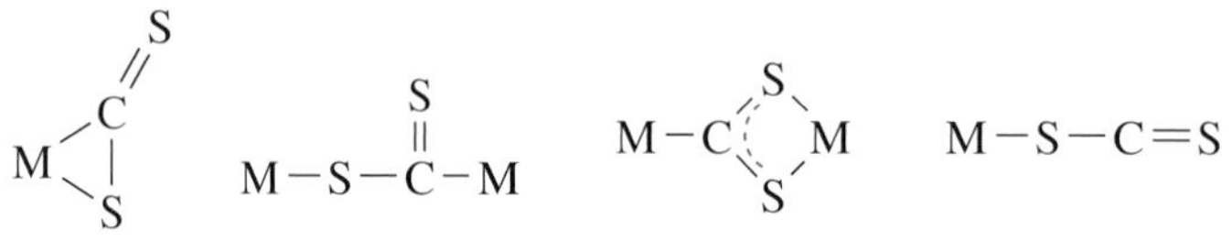

3.30 3.31 3.32 3.33

The heavier elements of Group 14 form bonds to transition metals, but, for silicon and germanium, examples are largely limited to ligands where the element is substituted with organic groups, for example: $[Ir(CO)(Cl)(H)(PPh_3)_2(SiMe_3)]$ (where Me represents the methyl group and Ph phenyl). Tin(II) compounds, such as $SnCl_3^-$ or $SnMe_2$, form a wide range of complexes with transition elements.

■ Can you think why tin(II) is an effective ligand?

□ Tin(II) compounds have a lone pair of electrons, and this feature is central to the formation of the tin–metal bond. The $SnCl_3^-$ anion can replace a variety of other ligands, such as Cl^-, CO and PPh_3, for example:

$$[PtCl_4]^{2-} + 2SnCl_3^- = [PtCl_2(SnCl_3)_2]^{2-} + 2Cl^-$$

3.4.3 Elements of Group 15

Nitrogen is one of the most common ligating atoms and is particularly important in biological chemistry. You have already met complexes with ligands based on the ammonia molecule. These ligands include the ammonia molecule itself (**3.34**), amines (**3.35**) and imines (**3.36**). They may also be chelating, as with 1,2-diaminoethane (en) (**3.37**). (The ligand was shown in Table 3.3.)

$M-NH_3$ $M-NH_2R$ $M=NR$ 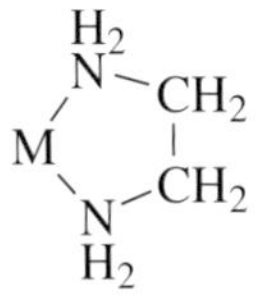

3.34 3.35 3.36 3.37

Nitrogen not only forms aliphatic organic compounds but also features in a range of aromatic molecules, many of which act as good ligands for transition metals. The 2,2′-bipyridyl (bipy) molecule (**3.38**), which you met in Table 3.3, can form bidentate complexes in which the metal can have several different oxidation states. In the series of chromium complexes $[Cr(bipy)_3]^{3+}$, $[Cr(bipy)_3]^{2+}$, $[Cr(bipy)_3]^{+}$, $[Cr(bipy)_3]$ and $[Cr(bipy)_3]^{-}$, the oxidation state of the metal varies in one unit steps from +3 to −1. The corresponding ammino complex, $[Cr(NH_3)_6]^{3+}$, with chromium in the +3 oxidation state is quite stable, but ammino complexes with chromium in lower oxidation states do not exist. This is because bipy can act as a π-acceptor, analogous to CO and CN^-.

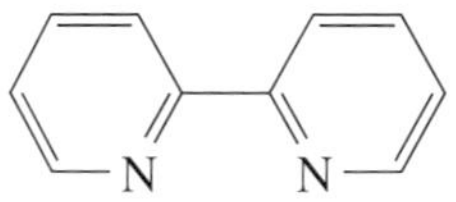

3.38

In addition to the substitution of hydrogen atoms in ammonia by organic groups, the molecule can be deprotonated to give the ions NH_2^-, NH^{2-} and N^{3-}, all of which can act as ligands.

Ligands with N–N bonds have been extensively studied as possible intermediates in the reduction of nitrogen, so-called **nitrogen fixation**, a very important process for fertiliser production. Table 3.6 gives an overview of some of the more important ligands.

Table 3.6 Ligands with N–N bonds, where R = H, alkyl or aryl group.

Ligand type	
hydrazine	$M–NR_2–NR_2$
hydrazido(1−)	$M–NH–NR_2$
diazenido	M–N=NR
hydrazido(2−)	$M=N–NR_2$

Porphyrin (or porphine) (structure **3.39**) is an example of a **tetrapyrrole**, an important type of naturally occurring macrocyclic ligand, which contains four nitrogen atoms held in a plane by a carbon framework: chlorophyll, haem and vitamin B_{12} contain metals bound by related ligands. In general, the metal lies in (or very close to) the plane of the four nitrogen atoms and is bound to each of them.

Phosphorus also forms macrocyclic ligands (**3.40**). One of the most important types of phosphorus ligand is the tertiary phosphine (PR_3) that is used extensively in the synthesis of complexes with the metal in a low oxidation state. Again, the bonding in these complexes is strengthened by the ability of phosphorus in these ligands to accept electrons from the metal. In addition to the monodentate PR_3 ligand, bidentate diphosphine ligands, such as $Ph_2PCH_2CH_2PPh_2$ (dppe; Table 3.5), are used extensively in metal-complex chemistry.

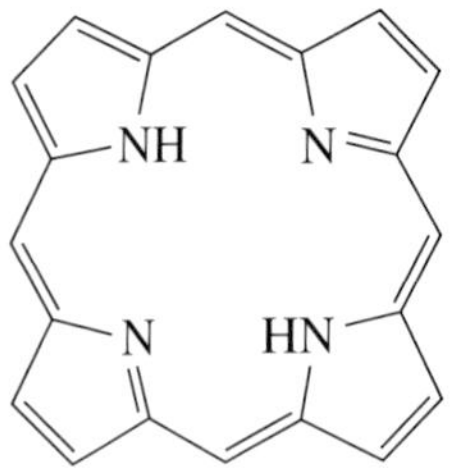

3.39

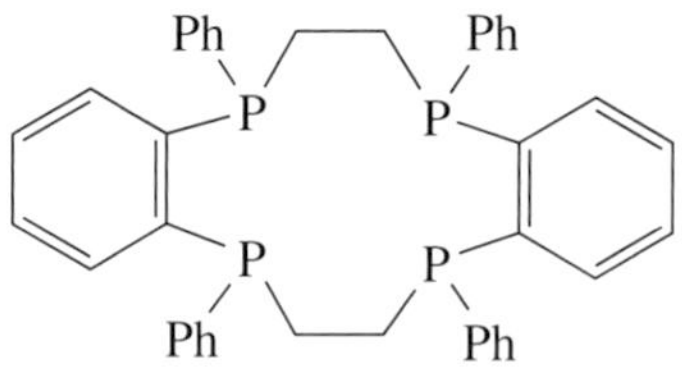

3.40

Compounds of arsenic, antimony and bismuth can act as ligands, although the latter two are very weak donors and, consequently, the chemistry of the complex is more limited.

3.4.4 Elements of Groups 16 and 17

Perhaps the most common example of oxygen acting as a ligand occurs in metal aquo complexes (where the water molecule is the ligand), such as $[Fe(H_2O)_6]^{2+}$, or hydroxo complexes, such as $[Co(H_2O)_5(OH)]^{2+}$. The water ligand is generally monodentate, but many examples are known where the hydroxide ligand can be bridging either across two metal atoms (μ_2) (**3.41**), or even across three metal atoms (μ_3) (**3.42**).

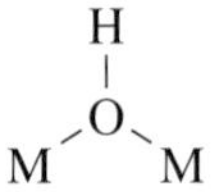

3.41

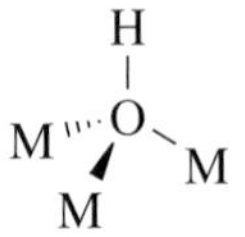

3.42

Ethers, ketones and esters can all act as oxygen ligands, and the crown ethers, a fascinating group of macrocyclic polyethers, have enabled a wide range of new complexes to be synthesised. The crown ethers (of which dibenzo-18-crown-6, **3.43**, is a common example) confer an unusual stability on complexes. This is further illustrated by the cryptand complexes of the alkali-metal cations. One such ligand is 2,2,2 cryptand (**3.44**), which has been used in the synthesis of compounds containing alkali-metal *anions* in which the corresponding metal cation is complexed by the cryptand, i.e. $[cryptand\text{–}Na]^+Na^-$. The metal cation can coordinate to the six oxygen atoms and the two nitrogen atoms, the size of the cavity in the ligand being critical for a particular metal ion.

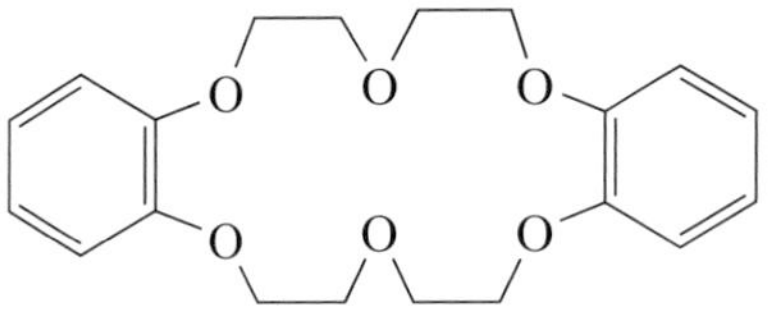

3.43

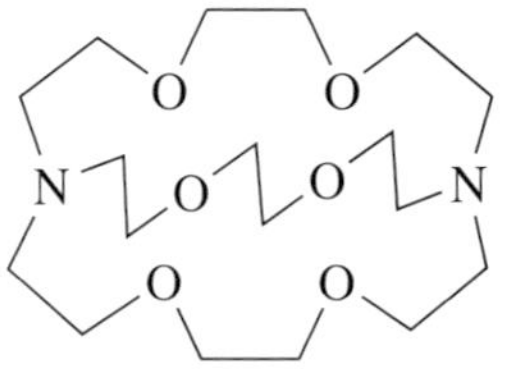

3.44

Many stable metal complexes are formed with β-diketonolate ions (**3.45**), a chelate which binds through the two oxygen atoms. The ions are produced by the removal of a proton after enolisation of the ketone.

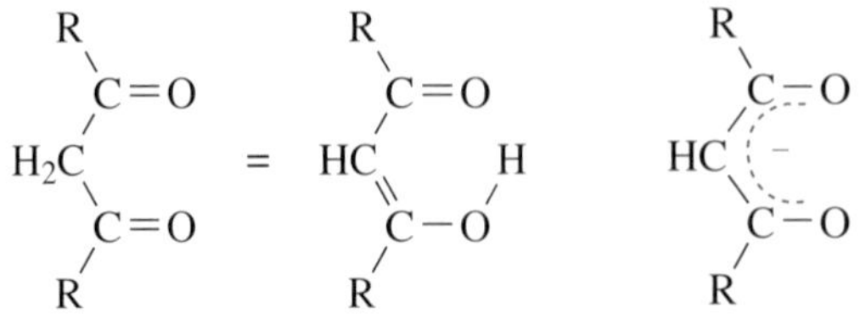

3.45

Another important class of oxygen ligands contains the oxoanions, the most common being CO_3^{2-}, NO_3^{-}, SO_3^{2-}, SO_4^{2-} and PO_4^{3-}. The carbonate anion (**3.46**) is usually unidentate, and the related oxalate dianion from oxalic acid is an effective bidentate ligand, typically forming trisoxalato complexes [M(ox)$_3$], as shown by structure **3.47**.

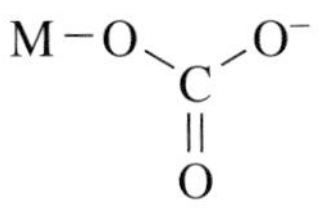

3.46

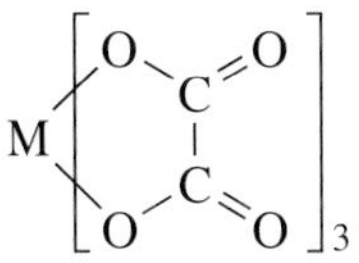

3.47

Many examples of oxygen ligands are paralleled in sulfur chemistry. The dithiocarbamate anion (**3.48**) is found in monodentate (**3.49**), chelating (**3.50**) and bridging (**3.51**) situations.

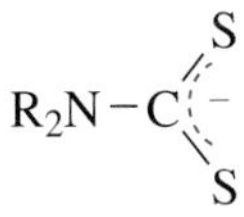

3.48

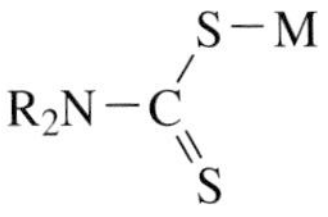

3.49

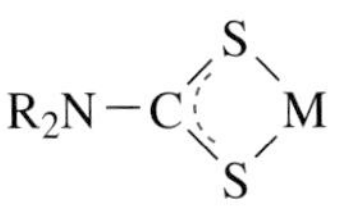

3.50

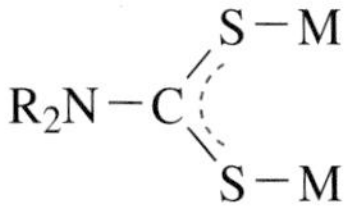

3.51

All the halide ions can act as monodentate ligands. In addition, they can form bridges in both main-Group and transition-metal complexes.

Don't forget that there are questions on the companion website which you can use to test your understanding of the material covered in this chapter.

4 The stability of coordination compounds

In this section the stability of metal complexes will be analysed in some detail from a thermodynamic standpoint. In particular we will consider the effect of changing ligands on the relative stability of different oxidation states for a given transition element.

As appropriate thermodynamic data are not always available, we will also introduce two empirical trends that have been identified. We will consider the stability of the complexes of a particular ligand as we move across the first transition metal series and also the observation that some ligands form their most stable complexes with species such as titanium(IV) and cobalt(III), whereas others are associated with species of the type Ag(I) and Pt(II).

4.1 Relative stability of metal complexes with varying ligands

So far in our discussion of thermodynamic stability (Chapter 2), we have considered one particular type of problem: the variation in the relative stability of two oxidation states when the transition element is varied and the ligands are fixed. Indeed throughout most of the discussion, the nature of the ligand has played a somewhat subsidiary role. We shall now focus on this role, and enquire about the effect that changing the ligands has on the relative stability of different oxidation states of a given transition element. As you saw in Activity 2.1, cobalt provides a particularly dramatic example. (You can access this video, *Relative stabilities of cobalt(II) and cobalt(III)*, from the video library on the DVD.)

Let us review the key points. $Co^{3+}(aq)$ is a powerful oxidising agent. Thus, oxidation of $Co^{2+}(aq)$ is very difficult, but possible through electrolysis in acid solution at 0 °C, which produces the deep blue colour of $Co^{3+}(aq)$. However, on standing at room temperature (or warming), $Co^{3+}(aq)$ steadily oxidises water: oxygen is evolved, and the blue colour of $Co^{3+}(aq)$ disappears to be replaced by the pink of $Co^{2+}(aq)$.

■ Are these observations consistent with the following standard electrode potentials?

$$Co^{3+}(aq) + e = Co^{2+}(aq) \quad E^{\ominus} = 1.94\ V \tag{4.1}$$

$$\tfrac{1}{2}O_2(g) + 2H^+(aq) + 2e = H_2O(l) \quad E^{\ominus} = 1.23\ V \tag{4.2}$$

□ Yes. $E^{\ominus}(Co^{3+}|Co^{2+})$ is greater (more positive) than $E^{\ominus}$ for the oxygen/water couple. Thus, in acid solution at 25 °C, $Co^{3+}(aq)$ is thermodynamically capable of oxidising water.

Put another way, electrode potentials tell us that the system represented by Equation 4.3 has $\Delta G_m^\ominus < 0$, so equilibrium lies over to the right.

$$2Co^{3+}(aq) + H_2O(l) = 2Co^{2+}(aq) + 2H^+(aq) + \tfrac{1}{2}O_2(g) \qquad (4.3)$$

Consider now the effect of adding ammonia to this system. As you may recall (Section 3.1), adding ammonia solution results in the formation of the ammino complex $[Co(NH_3)_6]^{2+}(aq)$, which has a similar (red) colour to the aquo complex, $[Co(H_2O)_6]^{2+}$. To emphasise the change of ligand, the complexing process can be represented by the equation:

$$[Co(H_2O)_6]^{2+}(aq) + 6NH_3(aq) = [Co(NH_3)_6]^{2+}(aq) + 6H_2O(l) \qquad (4.4)$$

Provided a little activated charcoal is added to speed things along, bubbling air or oxygen through the ammoniacal solution changes its colour to yellow-brown. Again, you may recall that the colour is due to the ammino complex of cobalt(III), $[Co(NH_3)_6]^{3+}(aq)$: the complexing process can be represented by an equation directly analogous to that for $Co^{2+}(aq)$ (Equation 4.4), as:

$$[Co(H_2O)_6]^{3+}(aq) + 6NH_3(aq) = [Co(NH_3)_6]^{3+}(aq) + 6H_2O(l) \qquad (4.5)$$

To draw these observations together, suppose the cobalt(III)/cobalt(II) system in Equation 4.3 is written in a more general form, as:

$$2Co(III)(aq) + H_2O(l) = 2Co(II)(aq) + 2H^+(aq) + \tfrac{1}{2}O_2(g) \qquad (4.6)$$

Evidently this equilibrium lies to the right when water is the ligand: in these circumstances, cobalt(II) is stable with respect to cobalt(III) in an oxygenated acid solution. But if ammonia is added, $[Co(NH_3)_6]^{2+}(aq)$ is formed, and oxidised by oxygen to $[Co(NH_3)_6]^{3+}(aq)$. In other words, the equilibrium in Equation 4.6 lies to the left when ammonia is the ligand. We conclude that changing the ligand from water to ammonia stabilises cobalt(III) with respect to cobalt(II).

This discussion is based (albeit implicitly) on an application of **Le Chatelier's principle**. We can argue that complex formation by $Co^{2+}(aq)$ with ammonia (Equation 4.4) effectively removes $Co^{2+}(aq)$ from the equilibrium system in Equation 4.6. This would tend to stabilise cobalt(II). The final outcome, the stabilisation of cobalt(III), must mean that 'removal' of $Co^{3+}(aq)$ by the analogous complexing reaction (Equation 4.5) is sufficient to offset this, hence 'tipping the balance' in Equation 4.6 over to the left.

4.1.1 Quantifying the argument – stability constants

Our next step places this conclusion – largely drawn from qualitative observations – on a proper quantitative footing.

■ What thermodynamic quantity characterises the equilibrium position in a reaction?

□ The equilibrium constant, usually denoted as K.

You may (with equal merit) have answered, 'the value of $\Delta G_{\mathrm{m}}^{\ominus}$ for the reaction': the two quantities are related as follows, where R (= 8.314 J K^{-1} mol^{-1}) is the gas constant

$$\Delta G_{\mathrm{m}}^{\ominus} = -2.303RT \log K \qquad (4.7)$$

The equilibrium constant for a complexing reaction, like Equation 4.4 or Equation 4.5, has a rather special significance: it provides a measure of the stability of the complex in question, with respect to the ligand and the corresponding aqueous ion. Not surprisingly, it is called the **stability constant** (or sometimes the **formation constant**) of the complex.

In your reading, you may find the symbol β (Greek beta) used to represent the stability constant of a complex as defined here.

By convention, the formation reaction is generally written without drawing any distinction between the six water molecules bound in the aqueous complex and the other water molecules in the bulk of the solution. Thus, Equation 4.4, for example, becomes:

$$Co^{2+}(aq) + 6NH_3(aq) = [Co(NH_3)_6]^{2+}(aq) \qquad (4.8)$$

This is an example of a **ligand exchange reaction**, where the water ligands in the inner sphere of the complex are exchanged for ammine ligands.

■ Write an expression for the stability constant of the cobalt(II) complex, $[Co(NH_3)_6]^{2+}(aq)$, in terms of the concentrations of the species in Equation 4.8.

□ The form of an equilibrium constant follows directly from the stoichiometry of the balanced reaction equation *as written*. From Equation 4.8 the stability constant of $[Co(NH_3)_6]^{2+}$ is given by:

$$K = \frac{[[Co(NH_3)_6]^{2+}]}{[Co^{2+}][NH_3]^6} \qquad (4.9)$$

where the symbol [X] is used to denote the concentration of species X.

Now look through our discussion of the effect of adding ammonia to the system in Equation 4.3:

$$2Co^{3+}(aq) + H_2O(l) = 2Co^{2+}(aq) + 2H^+(aq) + \tfrac{1}{2}O_2(g) \qquad (4.3)$$

■ What do you conclude about the relative sizes of the stability constants for the two complex ions, $[Co(NH_3)_6]^{3+}(aq)$ and $[Co(NH_3)_6]^{2+}(aq)$?

□ The expression in Equation 4.9 is typical of that for a stability constant. Evidently, the larger the stability constant for a given complex, the more effectively the ligand in question will 'mop up' the corresponding aqueous ion to form the complex. Given that ammonia stabilises cobalt(III) with respect to cobalt(II), the stability constant of the cobalt(III) complex must be greater than that of the cobalt(II) complex. In practice, it is greater by a factor of some 10^{31}.

The thrust of the discussion above is that the larger stability constant of $[Co(NH_3)_6]^{3+}(aq)$ stabilises cobalt in the +3 state. Now, if this is so, it should also be reflected in the electrode potential of the 'ammino' couple, $E^{\ominus}\left([Co(NH_3)_6]^{3+}|[Co(NH_3)_6]^{2+}\right)$: after all, it is this quantity that provides a measure of the 'oxidising power' of a given redox couple.

■ Do the following values of $E^{\ominus}$ support the conclusion drawn above?

$$Co^{3+}(aq) + e = Co^{2+}(aq) \quad E^{\ominus} = 1.94\ V \tag{4.1}$$

$$[Co(NH_3)_6]^{3+}(aq) + e = [Co(NH_3)_6]^{2+}(aq) \quad E^{\ominus} = 0.1\ V \tag{4.10}$$

□ Yes. We have argued that the more positive the value of $E^{\ominus}(M^{3+}|M^{2+})$, the more stable is M^{2+} with respect to M^{3+}. Thus, the lowering of $E^{\ominus}$ (by 1.84 V) represents a substantial stabilisation of the +3 state.

So we now appear to have *two* descriptions of the stabilisation of cobalt(III) – the one in terms of Le Chatelier's principle and the stability constants of the ammino complexes, and the other in terms of the lowering of the electrode potential. Since both descriptions involve thermodynamic quantities, we might expect them to be connected in some way. As you will see in the next section, the Nernst equation provides this link.

4.1.2 The Nernst equation

Consider again Reaction 4.1

$$Co^{3+}(aq) + e = Co^{2+}(aq) \quad E^{\ominus} = 1.94\ V \tag{4.1}$$

A crucial point to note is that, strictly speaking, values of $E^{\ominus}$ (like the equivalent values of $\Delta G_m^{\ominus}$) afford a comparison of the oxidising power of redox couples when, and only when, all the species in the redox equation are in their *standard* states (exemplified by the symbol $\ominus$). Thus, the value of 1.94 V is a thermodynamic measure of the 'strength' of $Co^{3+}(aq)$ as an oxidising agent when both $Co^{2+}(aq)$ and $Co^{3+}(aq)$ are in their standard states. The standard state, for an ionic solute in aqueous solution, like $Co^{2+}(aq)$ or $Co^{3+}(aq)$ for example, can be taken as equivalent to a *unit concentration*, 1 mol dm^{-3}.

(*Note*: Strictly speaking the standard state is taken as unit activity, *a*, rather than concentration, *c*; they are related by, $a = c/c^{\ominus}$, where $c^{\ominus}$ is the standard concentration. Throughout this book we shall adopt the standard concentration, $c^{\ominus} = 1\ \text{mol dm}^{-3}$, such that the standard state is established at a concentration $c = c^{\ominus} = 1\ \text{mol dm}^{-3}$.)

Recall the $E^{\ominus}$ values for both the aquo and the ammino couples:

$$Co^{3+}(aq) + e = Co^{2+}(aq) \quad E^{\ominus} = 1.94\ V \tag{4.1}$$

$$[Co(NH_3)_6]^{3+}(aq) + e = [Co(NH_3)_6]^{2+}(aq) \quad E^{\ominus} = 0.1\ V \tag{4.10}$$

Now, in line with the discussion above, 0.1 V can obviously be regarded as the standard potential of the ammino couple, appropriate to a solution containing both complex ions, each at unit concentration. But such a solution must also contain finite (albeit very small) concentrations of the two aqueous (aquo complex) ions, $Co^{2+}(aq)$ and $Co^{3+}(aq)$, according to the equilibria described by the stability constants of the two complexes. This, in turn, suggests an alternative interpretation of the value 0.1 V, namely that it represents the potential of the ($Co^{3+}(aq) \mid Co^{2+}(aq)$) couple, but under conditions markedly different from those to which the standard value (1.94 V) applies. From this point of view, then, the effect of complex formation is to 'drive' the system in Equation 4.1 away from its standard state. In these circumstances, the quantity that provides a measure of the oxidising power of the aquo couple is not $E^{\ominus}$, but E – where dropping the superscript indicates that we are no longer working at unit concentration and standard conditions. The link between E and $E^{\ominus}$ for the couple between the aquo ions (and hence $E^{\ominus}$ for the ammino complex) is provided by the **Nernst equation**.

According to the Nernst equation, the electrode potential for the aquo couple (Equation 4.1) under these non-standard conditions is determined by:

$$E(Co^{3+}(aq)|Co^{2+}(aq)) = E^{\ominus}(Co^{3+}(aq)|Co^{2+}(aq)) - (0.0592\ V)\log\left(\frac{[Co^{2+}]}{[Co^{3+}]}\right) \tag{4.11}$$

where $[Co^{2+}]$ and $[Co^{3+}]$ are the concentrations of $Co^{2+}(aq)$ and $Co^{3+}(aq)$ ions, respectively, in the solution containing unit concentrations of the corresponding complex ions, $[Co(NH_3)_6]^{2+}(aq)$ and $[Co(NH_3)_6]^{3+}(aq)$. Thus,

$$\begin{aligned} E &= E^{\ominus} - (0.0592\ V)\log\left(\frac{[Co^{2+}]}{[Co^{3+}]}\right) \\ &= 1.94\ V - (0.0592\ V)\log\left(\frac{[Co^{2+}]}{[Co^{3+}]}\right) \end{aligned} \tag{4.12}$$

■ Now use the stability constant to write an expression for the concentration of $Co^{3+}(aq)$ in a solution containing the complex $[Co(NH_3)_6]^{3+}(aq)$ at unit concentration (1 mol dm^{-3}).

□ Using the symbol $K^{\ominus}_{+3}$ to indicate the stability constant for $[Co(NH_3)_6]^{3+}(aq)$:

$$K^{\ominus}_{+3} = \frac{[[Co(NH_3)_6]^{3+}]}{[Co^{3+}][NH_3]^6}$$

Thus $[Co^{3+}] = \dfrac{[[Co(NH_3)_6]^{3+}]}{K^{\ominus}_{+3}[NH_3]^6}$

With $[[Co(NH_3)_6]^{3+}] = 1$

$$[Co^{3+}] = \frac{1}{K^{\ominus}_{+3}\,[NH_3]^6} \tag{4.13}$$

Likewise, the stability constant of the cobalt(II) complex can be written $K^{\ominus}_{+2}$, such that

$$[Co^{2+}] = \frac{1}{K^{\ominus}_{+2}\,[NH_3]^6} \tag{4.14}$$

Combining these results and substituting into Equation 4.12 gives:

$$E = 1.94\ \text{V} - (0.0592\ \text{V})\log\left(\frac{K^{\ominus}_{+3}}{K^{\ominus}_{+2}}\right)$$

Recalling from Section 4.1.1 that $K^{\ominus}_{+3}/K^{\ominus}_{+2} = 10^{31}$, this becomes

$$E = 1.94\ \text{V} - (0.0592\ \text{V})\ \log\ (10^{31}) = 1.94\ \text{V} - 1.84\ \text{V} = 0.10\ \text{V}$$

in agreement with the value of $E^{\ominus}$ for the ammino couple quoted earlier.

There are several points to note here. First, the analysis outlined above is based on completely general principles: in particular, it can be applied whenever a metal forms both aquo ions in two different oxidation states, and corresponding complex ions with a given ligand. Provided the stability constants of the complex ions are known, the standard potential of the 'complex' couple – call it $E^{\ominus}_{\text{comp}}$, say – can always be calculated from that of the aquo couple, as in the example above and shown as a general case for dipositive and tripositive ions in Box 4.1.

Because of this link, $E^{\ominus}$ of the aquo couple provides a kind of baseline against which to measure the ability of a given ligand to stabilise one oxidation state with respect to the other. Thus, the calculation embodied in Equation 4.12 is a concrete expression of the dramatic stabilisation of the +3 state of cobalt by ammonia.

This highlights a second, related point: the stability constants of complexes may differ by many orders of magnitude – a factor of 10^{31} in our example. Thus, despite the logarithmic dependence in the Nernst equation, the stability of a given oxidation state may, as you have seen, change very markedly on the addition of a ligand that forms stable complexes. Indeed, this chemistry is utilised in nature in the uptake of iron by bacteria from soil. Iron(III) forms very stable complexes with a particular class of ligand known as a **siderophore**, which the bacteria secrete into the soil for this very purpose.

Box 4.1 Nernst equation

Consider the general couple

$$[ML_6]^{3+}(aq) + e = [ML_6]^{2+}(aq) \tag{4.15}$$

for a metal M that exists as aqueous ions in oxidation states, $M^{2+}(aq)$ and $M^{3+}(aq)$, and forms complexes $[ML_6]^{2+}(aq)$ and $[ML_6]^{3+}(aq)$ with a ligand, L. The standard potential of the couple in a solution containing *unit concentrations of these complex ions* is measured by $E^{\ominus}_{comp}$. This will equate to E for the corresponding aquo couple such that

$$\begin{aligned} E^{\ominus}_{comp} &= E(M^{3+}(aq)|M^{2+}(aq)) \\ &= E^{\ominus}(M^{3+}(aq)|M^{2+}(aq)) - (0.0592\ V)\log\left(\frac{[M^{2+}]}{[M^{3+}]}\right) \end{aligned} \tag{4.16}$$

where $[M^{2+}]$ and $[M^{3+}]$ are the concentrations of $M^{2+}(aq)$ and $M^{3+}(aq)$ ions, respectively, in the solution containing unit concentrations of the corresponding complex ions, $[ML_6]^{2+}(aq)$ and $[ML_6]^{3+}(aq)$.

This is equivalent to

$$E^{\ominus}_{comp} = E^{\ominus}(M^{3+}(aq)|M^{2+}(aq)) - (0.0592\ V)\log\left(\frac{K^{\ominus}_{+3}}{K^{\ominus}_{+2}}\right) \tag{4.17}$$

where $K^{\ominus}_{+2}$ and $K^{\ominus}_{+3}$ are the stability constants of the corresponding complex ions, $[ML_6]^{2+}(aq)$ and $[ML_6]^{3+}(aq)$, respectively.

Please note that for the purposes of this text, it is important only that you recognise the Nernst equation – and know how to use it to analyse and discuss the sort of chemical problem outlined above. To this end, we have not attempted to derive the equation here, nor explore its thermodynamic 'roots', and have simply quoted the final result.

4.1.3 The pH-dependence of electrode potentials – 'solvent decomposition'

Another important factor in any aqueous solution is the concentration of hydrogen ions. Since this may vary by a factor of around 10^{14} in water, a change in pH may cause a marked alteration in the behaviour of particular oxidation states, even where it does not change the nature of the complexes involved.

In this context, as we have seen in Section 2, there are two couples which are particularly important in aqueous solution:

The first, the oxygen electrode

$$\tfrac{1}{2}O_2(g) + 2H^+(aq) + 2e = H_2O(l) \quad E^{\ominus} = 1.23\ V \tag{4.2}$$

defines both the resistance of water to oxidation, and the 'oxidising power' of atmospheric oxygen in contact with an aqueous solution.

The second, the hydrogen electrode

$$H^+(aq) + e = \frac{1}{2}H_2(g) \quad E^{\ominus} = 0\ V \tag{4.18}$$

defines the resistance to reduction of the hydrogen ions in an aqueous solution.

Using the Nernst equation, these have been found to vary with pH according to the equations below:

$$E/V = 1.23 - 0.0592\ pH \quad \text{'oxygen electrode'} \tag{4.19}$$

$$E/V = -\ 0.0592\ pH \quad \text{'hydrogen electrode'} \tag{4.20}$$

■ Why is the potential in Equations 4.19 and 4.20 expressed as E not $E^{\ominus}$?

□ The oxygen and hydrogen electrodes are no longer operating under standard conditions.

Thus, the Nernst equation allows us to extend the analysis of aqueous redox chemistry to neutral and alkaline solutions. In this context, the expressions in Equations 4.19 and 4.20 can be seen effectively to limit the thermodynamic stability of oxidising or reducing agents in aqueous solution to couples whose potentials lie in the range (−0.0592 pH) V to (1.23 − 0.0592 pH) V.

Finally, we close this section by returning to the cobalt–ammino system. Consider again Reactions 4.1 and 4.10. The standard potential of the aquo couple (1.94 V) is well above the value of 1.23 V required for the oxidation of water in acid solution, and so at room temperature, $Co^{2+}(aq)$ is favoured.

Adding ammonia produces two effects. Firstly, the nature of the cobalt species changes: ammino complexes are formed, and the potential of the Co(III)/Co(II) couple plummets to 0.1 V. Secondly, the pH of the solution will increase as aqueous ammonia is a weak base by virtue of the reaction:

$$NH_3(aq) + H_2O(l) = NH_4^+(aq) + OH^-(aq) \tag{4.21}$$

A typical value of the pH in the final solution is about 8.

■ What is the potential of the oxygen electrode in these circumstances?

□ In these circumstances, the potential of the oxygen electrode is $E = (1.23 - 0.0592 \times 8)\ V = 0.76\ V$.

Since this is larger (more positive) than the potential of the cobalt–ammino couple, oxygen in contact with an aqueous solution of pH 8 is thermodynamically capable of oxidising $[Co(NH_3)_6]^{2+}(aq)$ to the tripositive state, which indeed it does!

4.2 Variation in stability constants of complexes across the transition series

Recall the ligand exchange reaction in Reaction 4.8. This is shown in more general terms (in this case for dipostive ions) as:

$$M^{2+}(aq) + nL^{x-}(aq) = [ML_n]^{(2-nx)+}(aq) \quad (4.22)$$

where water is exchanged with another ligand L^{x-}. As we saw in the previous section, the equilibrium constant, $K^{\ominus}$, is called the stability constant of the complex, where

$$K^{\ominus} = \frac{[[ML_n]^{(2-nx)+}(aq)]}{[M^{2+}(aq)][L^{x-}(aq)]^n}$$

The larger the value of $K^{\ominus}$, the more stable is the complex.

The values of the stability constant for the ligands oxalate, $C_2O_4^-$; glycine, $H_2NCH_2CO_2^-$; and en, $H_2NCH_2CH_2NH_2$ across the first transition series (and beyond) of metal ions is illustrated in Figure 4.1 for the dipositive ions. Although the values vary for each ligand, we can see that the relative order across the series is largely independent of the ligand such that:

$$Ba^{2+} < Sr^{2+} < Ca^{2+} < Mg^{2+} < Mn^{2+} < Fe^{2+} < Co^{2+} < Ni^{2+} < Cu^{2+} > Zn^{2+}$$

This sequence is called the **Irving–Williams order**.

4.3 Hard and soft acids and bases

You may have noted in Figure 4.1 that some metal ions form particularly stable complexes with particular ligands. For example, of the three ligands illustrated, the ligand en forms the most stable complex with Cu^{2+} and the least stable with Mg^{2+}. Metal ions and ligands can be classified as **hard and soft acids and bases** as in Table 4.1 reflecting their Lewis acid/base ability.

Table 4.1 Hard and soft acids and bases.

	Hard	Borderline	Soft
Acids	H^+, Li^+, Na^+, K^+, Be^{2+}, Mg^{2+}, Ca^{2+}, BF_3, BCl_3, $B(OR)_3$, Al^{3+}, Mn^{2+}, $AlCl_3$, $Al(CH_3)_3$, Sc^{3+}, Ti^{4+}, VO^{2+}, Cr^{3+}, Fe^{3+}, Co^{3+}, La^{3+}	Fe^{2+}, Co^{2+}, Ni^{2+}, Cu^{2+}, Zn^{2+}, Rh^{3+}, $B(CH_3)_3$, R_3C^+, Sn^{2+}, Pb^{2+}	Cu^+, Ag^+, Au^+, Cd^{2+}, Hg^{2+}, Pt^{2+}, Pt^{4+}, Pd^{2+}, CH_3Hg^+, Tl^+
Bases	NH_3, RNH_2, N_2H_4, H_2O, OH^-, O^{2-}, ROH, RO^-, CO_3^{2-}, SO_4^{2-}, ClO_4^-, F^-, Cl^-	$PhNH_2$, N_3^-, N_2, Br^-	H^-, R^-, C_2H_4, C_6H_6, CN^-, CO, SCN^-, R_3P, R_2S, RSH, RS^-, S^{2-}, I^-

Hard acids include the cations of the alkali metals, the alkaline earths, the lanthanides and the first-row transition metals in high oxidation states. They occur among the metallic elements to the left of the Periodic Table. The soft acids such as Ag^+, Cd^{2+} and Hg^{2+} are ions of metallic elements towards the right.

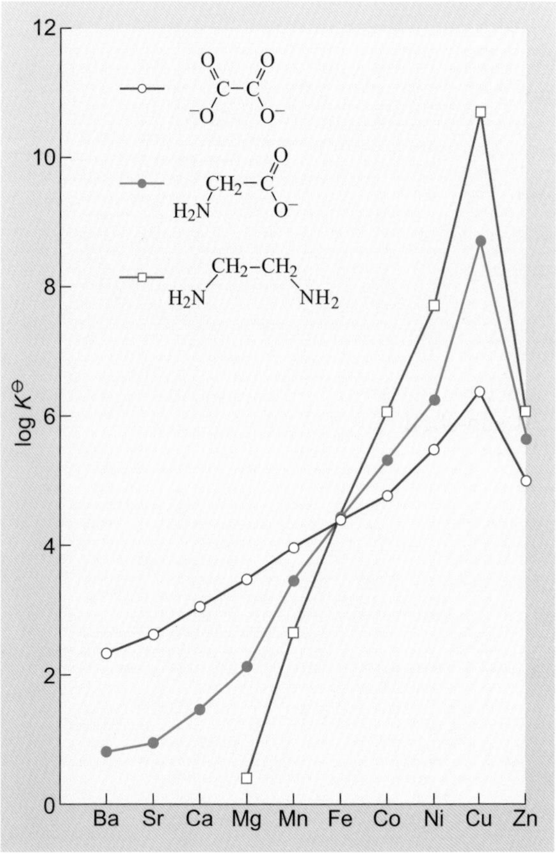

Figure 4.1 The Irving–Williams series for three different ligands, oxalate, glycine and en. ($K^\ominus$ is the equilibrium constant for Reaction 4.22.)

Hard bases include the fluoride anion and ligands that donate through oxygen. All oxygen donors in Table 4.1 are classified as hard. Nitrogen donors are less hard: some like ammonia are placed in the hard category, but others are borderline. Hard bases tend to have donor atoms with low polarisability. As one moves from nitrogen to fluorine, electrons are held more tightly so the electron distribution is less easily distorted by an external charge or electric field and hardness increases. However, as one descends the halogen group, electrons become more numerous and less tightly bound, so now it is softness that increases. Thus in Table 4.1, I^- is classified as soft, Cl^- and Br^- as borderline, and F^- as hard. Likewise, S^{2-} is soft and O^{2-} is hard.

These ideas are useful because of the simple observation that *hard acids tend to bind to hard bases and soft acids to soft bases.* For example, the metals calcium, aluminium, silver and mercury occur naturally as ores of calcium(II), aluminium(III), silver(I) and mercury(II). The donor atoms in metallic ores are usually oxygen or sulfur. Table 4.1 implies that the hard acids Ca^{2+} and Al^{3+} will bind to hard oxygen donors; the soft acids, Ag^+ and Hg^{2+} to soft sulfur donors. This is correct: aluminium is found as the oxide ore, bauxite, and calcium as the carbonate; both mercury and silver occur naturally as sulfide ores.

The influence of hardness and softness can also be detected in Figure 4.1. Superimposed on the general increase in the stability of the complex with respect to its dissociation into components is a hardness and softness effect in which the softer metals, coming later in the series (greater number of d electrons, so more polarisable), favour the softer ligands; so en (N-donating) is softer than oxalate (O-donating) and thus forms the most stable complex with the soft metal (Cu^{2+}).

- ■ Does the trend, observed in Figure 4.1, in the stability of the complexes for Mg^{2+} reflect the hardness of the different ligands?

- □ Yes. Mg^{2+} is a hard acid, and so it forms the most stable complex with the harder O-donating oxalate ligand. The softer N-donating en ligand forms the least stable complex.

Don't forget that there are questions on the companion website which you can use to test your understanding of the material covered in this chapter.

Please note that this hard/soft acid–base theory should only be used as an aid to *predict the preference* of a specific ligand with particular metals. A 'hard' acid, for example, may also form a complex with a 'soft' base, as in the example of Fe^{3+} which will form a complex with the S containing amino acid cysteine.

5 Crystal-field theory

In Chapter 1, we mentioned that one of the most striking properties of transition-metal complexes is the wide range of colours they exhibit. There are also intriguing variations in the magnetic behaviour of transition-metal complexes. For example, although they both contain central Fe^{2+} ions, $[Fe(H_2O)_6]^{2+}$ is paramagnetic (it is attracted into a magnetic field), but $[Fe(CN)_6]^{4-}$ is diamagnetic (it is weakly repelled by a magnetic field). In this chapter, we shall look at a bonding theory to help us explain these, and other, observations.

The key attribute of transition-metal ions which gives rise to many of these properties is their possession of partially occupied d orbitals. Across the fourth row of the Periodic Table, an electron enters the 4s sub-shell at potassium, and a second fills it at calcium. Then, from scandium to zinc, the 3d sub-shell is progressively filled. For the neutral atoms, the energies of the 3d and 4s orbitals are very close, and it is the stabilisation associated with half-filled and filled shells that gives rise to configuration irregularities at chromium and copper, respectively. This is shown in Table 5.1, where [Ar] represents the argon core electrons.

Table 5.1 Electronic configurations of the free atoms, dipositive ions and tripositive ions of the first transition series.

Element	Free atom configuration	M^{2+}	Configuration	M^{3+}	Configuration
Sc	$[Ar]3d^14s^2$	Sc^{2+}	$[Ar]3d^1$ *	Sc^{3+}	$[Ar]3d^0$
Ti	$[Ar]3d^24s^2$	Ti^{2+}	$[Ar]3d^2$	Ti^{3+}	$[Ar]3d^1$
V	$[Ar]3d^24s^2$	V^{2+}	$[Ar]3d^3$	V^{3+}	$[Ar]3d^2$
Cr	$[Ar]3d^54s^1$	Cr^{2+}	$[Ar]3d^4$	Cr^{3+}	$[Ar]3d^3$
Mn	$[Ar]3d^54s^2$	Mn^{2+}	$[Ar]3d^5$	Mn^{3+}	$[Ar]3d^4$
Fe	$[Ar]3d^64s^2$	Fe^{2+}	$[Ar]3d^6$	Fe^{3+}	$[Ar]3d^5$
Co	$[Ar]3d^74s^2$	Co^{2+}	$[Ar]3d^7$	Co^{3+}	$[Ar]3d^6$
Ni	$[Ar]3d^84s^2$	Ni^{2+}	$[Ar]3d^8$	Ni^{3+}	$[Ar]3d^7$
Cu	$[Ar]3d^{10}4s^1$	Cu^{2+}	$[Ar]3d^9$	Cu^{3+}	$[Ar]3d^8$
Zn	$[Ar]3d^{10}4s^2$	Zn^{2+}	$[Ar]3d^{10}$	Zn^{3+}	– *

* Compounds of scandium(II) are very rare and those of Zn(III) are non-existent.

When transition-metal atoms form cations, the 4s electrons are lost first. On ionisation, the 3d orbitals are significantly more stabilised (that is, drop to lower energy) than the 4s would be. This stems from the fact that the 3d electrons are not shielded from the nucleus as well as the 4s electrons. Therefore, the +2 and +3 ions have electronic configurations of $[Ar]3d^n$ (or $1s^22s^22p^63s^23p^63d^n$). The electronic configurations of the +2 and +3 ions, which we shall refer to frequently, are also shown in Table 5.1.

5.1 The d orbitals

There are five d orbitals, which, with reference to a set of mutually perpendicular axes, can be represented by their **boundary surfaces**, the contours inside which a d electron is found 95% of the time. The orbitals shown in Figure 5.1 are strictly those for an electron in a hydrogen atom, but those for electrons in transition-metal ions have the same shape. Note that the 3d orbitals are all presented from the same perspective, such that the *xz*-plane is always the plane of the paper.

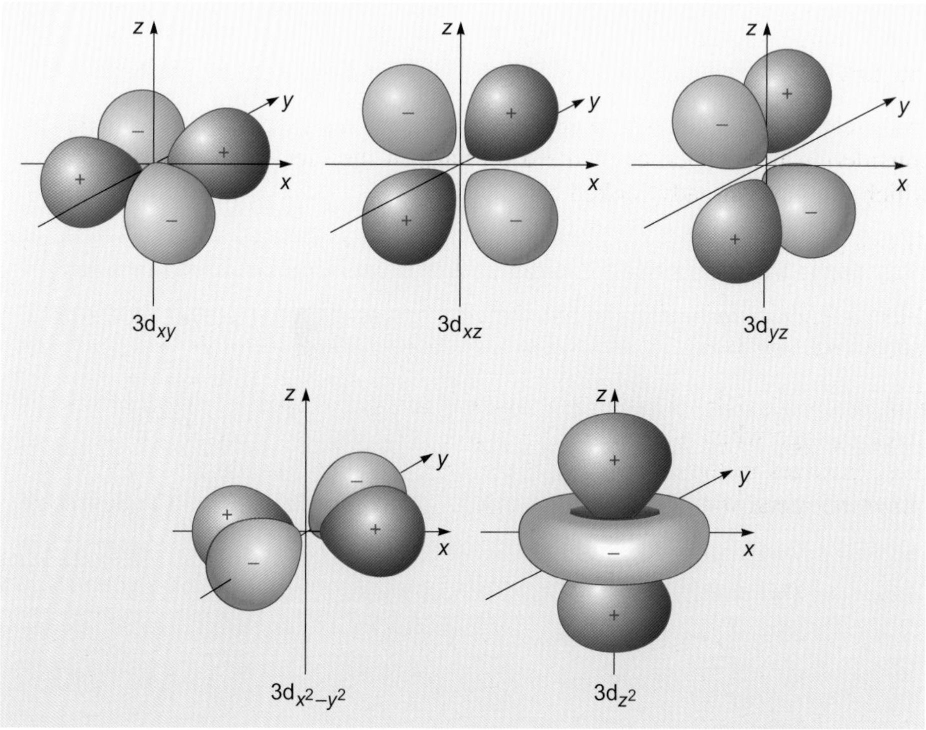

Figure 5.1 The shapes and orientation of the 3d orbitals, all shown in the *xz*-plane.

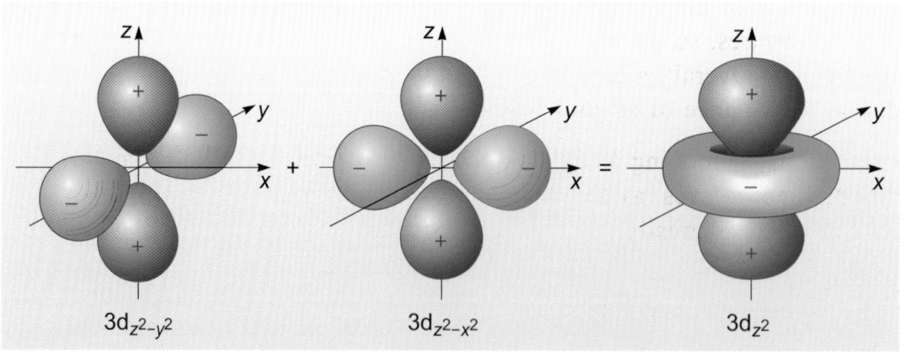

Figure 5.2 The components of $3d_{z^2}$: the $3d_{z^2-y^2}$ and $3d_{z^2-x^2}$ orbitals.

Four of these orbitals have the same shape but are orientated in different directions: the lobes of the $3d_{xy}$, $3d_{yz}$ and $3d_{xz}$ orbitals are *between* the relevant coordinate axes, whereas the lobes of the $3d_{x^2-y^2}$ orbital are *along* the x- and y-axes. The fifth, $3d_{z^2}$, is a combination of two orbitals, $3d_{z^2-y^2}$ and $3d_{z^2-x^2}$, which are shaped like the other four (Figure 5.2).

The energies of electrons in the 3d orbitals in transition metals are lower than that of an electron in a 3d orbital in the hydrogen atom (3d ionisation energies for the first transition series metals are in the range $12.8–17.6 \times 10^{-19}$ J as opposed to 2.4×10^{-19} J for hydrogen). This is because the 3d electrons in transition metals experience a greater nuclear charge than in hydrogen. As a consequence, the boundary surfaces of transition-metal 3d orbitals lie closer to the nucleus than those in hydrogen.

In developing a theory of bonding in transition-metal complexes, we start by considering the properties of the d orbitals on the metal ion, the shapes of which we have already looked at.

First we introduce **crystal-field theory**. This simple approach provides us with a remarkable insight into the chemical and physical properties of complexes of d-block metals. In Chapter 6, we develop a second, more comprehensive theory, molecular orbital theory, for the cases where this model is inadequate.

The key to crystal-field theory is that the energy of the 3d orbitals on the metal change when going from the free ion or atom to a metal complex. It is these changes in the energies of the d orbitals when ligands are added to a transition-metal ion that concern us here.

5.2 The basis of crystal-field theory

Crystal-field theory assumes that ligands behave as point negative charges, and the theory refers to the electric field produced by these charges. The nuclei, electrons and ligands interact purely through electrostatic forces. Firstly, there is an attraction between the negatively charged ligands and the positively charged metal ion. Secondly, and in opposition to this, the electrons in the metal-ion d orbitals increase in energy because of repulsion by the (negatively charged) ligands, so the energy of the metal-ion d orbitals rises. The magnitude of the electrostatic interaction depends on the distance between the charge centres, so we also need to consider how close the d electrons are to the ligands. Overall, a complex is stabilised relative to the free ion, and this is due to the balance of all these factors.

We shall start by looking at the application of crystal-field theory to **octahedral complexes**, as this geometry is one of the most common in transition-metal chemistry.

5.3 Octahedral complexes

For a free ion, the 3d orbitals are energetically equivalent (as shown in Figure 5.3a); they are referred to as being **degenerate**. We have seen that the energy of the d orbitals increases because of repulsion by negative charges. If there were a sphere of negative charges surrounding the metal ion, then all five energy levels will be raised equally (Figure 5.3b). In crystal-field theory, we assume that the negative charges representing the ligands are concentrated at six points representing six octahedrally arranged ligands, two are on the *x*-axis, two on the *y*-axis and two on the *z*-axis (Figure 5.4). We just noted that the energies of all the d orbitals increase, but the key question is: are all the 3d orbitals equally affected by this charge?

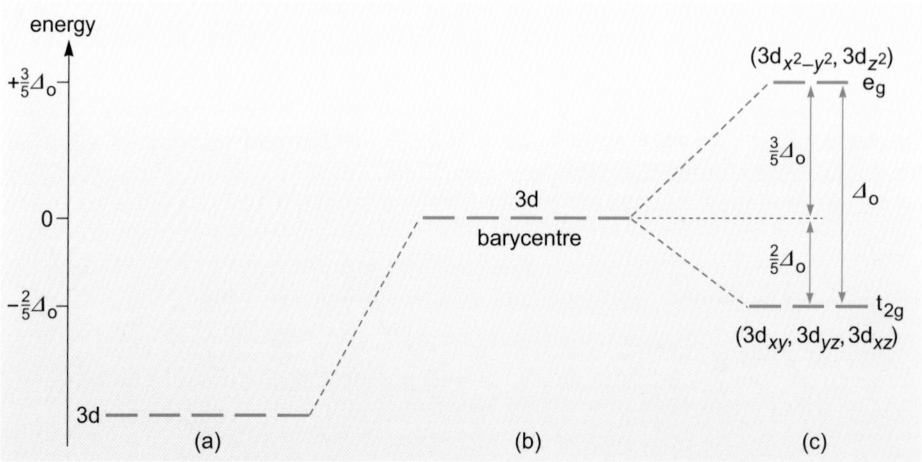

Figure 5.3 Partial orbital energy-level diagram (showing 3d levels only) for: (a) a free transition-metal ion; (b) a transition-metal ion in a sphere of negative charge (note that the d orbitals are raised in energy compared to the free ion); and (c) a transition-metal ion in an octahedron of six point negative charges.

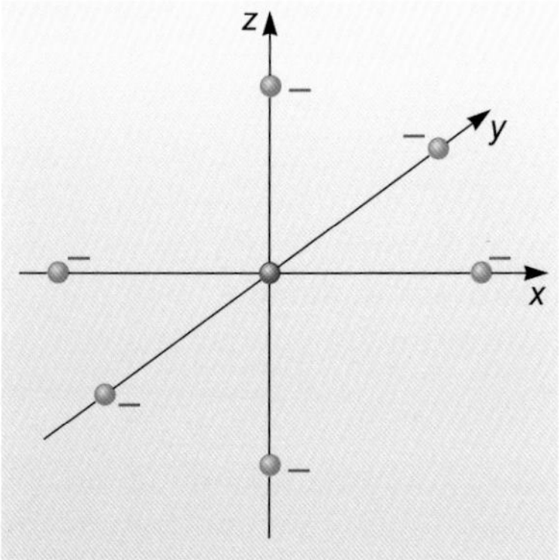

Figure 5.4 Six octahedrally disposed ligands represented as point negative charges.

Look first at the two d orbitals that are orientated in the *xy*-plane. Figure 5.5 shows the $3d_{xy}$ and $3d_{x^2-y^2}$ orbitals in the *xy*-plane, and the point charges on the *x*- and *y*-axes. By taking this bird's-eye view down the *z*-axis, you can see that whereas the $3d_{x^2-y^2}$ lobes are concentrated towards the point charges, those of the $3d_{xy}$ orbital lie between the charges. An electron in a $3d_{x^2-y^2}$ orbital comes closer to the point charges on average, than does an electron in a $3d_{xy}$ orbital. Thus, the $3d_{x^2-y^2}$ electron is repelled more by the ligands, and hence the $3d_{x^2-y^2}$ orbital is *higher* in energy than the $3d_{xy}$ orbital.

Similarly, if you look at the *xz*-plane, you will find that an electron in $3d_{z^2}$ experiences a greater repulsion than one in the $3d_{xz}$ orbital. Likewise, if you considered the *yz*-plane, you would find that an electron in $3d_{z^2}$ would be repelled more than an electron in the $3d_{yz}$ orbital.

To summarise: for a set of octahedrally arranged charges (an octahedral crystal field), the energy of the orbitals aligned along the axes ($3d_{x^2-y^2}$ and $3d_{z^2}$) is higher than those of the $3d_{xy}$, $3d_{xz}$ and $3d_{yz}$ orbitals, which are aligned between the axes (that is, further away from the ligands). This is represented in the form of the energy-level diagram in Figure 5.3c.

There are several points to note about this diagram. In both Figure 5.3a and 5.3b, the five d orbitals all have the same energy. The energy level in Figure 5.3b represents the average energy of the orbitals in the complex, known as the **barycentre**. This level corresponds to the hypothetical situation in which the metal ion was surrounded by a sphere of negative charge. The

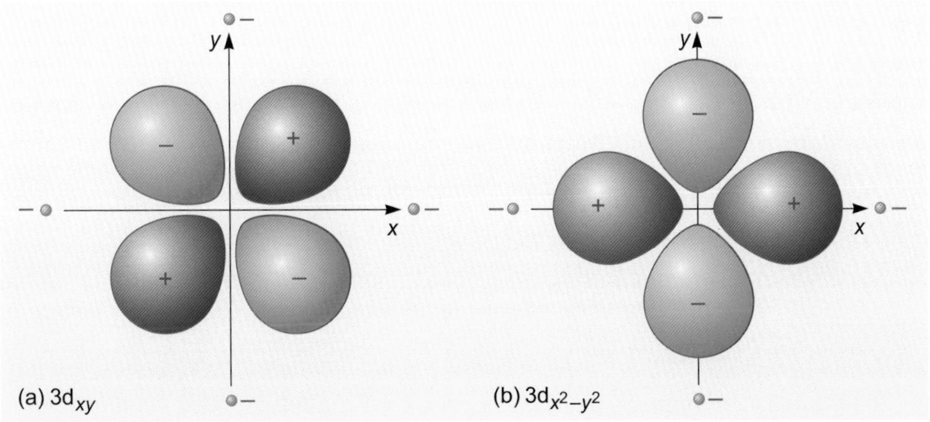

Figure 5.5 (a) 3d$_{xy}$ and (b) 3d$_{x^2-y^2}$ orbitals surrounded by point negative charges.

splitting of the orbitals in an octahedral crystal field is shown in Figure 5.3c; they are 'balanced' about the barycentre. Further, in an octahedral complex, the 3d$_{xy}$, 3d$_{xz}$ and 3d$_{yz}$ orbitals are degenerate (energetically equivalent), as are the 3d$_{z^2}$ and 3d$_{x^2-y^2}$ orbitals. The symbol Δ_o (pronounced 'delta oh': 'delta' to signify a difference in energy and 'oh' for octahedral) denotes the energy separation between the two sets of orbitals, and is referred to as the **crystal-field splitting energy**. The 3d$_{xy}$, 3d$_{xz}$ and 3d$_{yz}$ orbitals have an energy $\frac{2}{5}\Delta_o$ less than the average energy of the orbitals, and the 3d$_{z^2}$ and 3d$_{x^2-y^2}$ orbitals are raised $\frac{3}{5}\Delta_o$ higher than the average. (Note that when the five orbitals each contain one electron, the energy of the electrons in the three orbitals below the barycentre ($3 \times \frac{2}{5}\Delta_o$) is exactly balanced by the energy of the electrons in the two orbitals above ($2 \times \frac{3}{5}\Delta_o$).)

Box 5.1 Orbital labels

The levels in Figure 5.3 are labelled $\mathbf{t_{2g}}$ and $\mathbf{e_g}$. These are examples of a labelling system used for orbitals in molecules containing three or more atoms. A single orbital with a particular energy is labelled a or b. Two energy levels with the same energy are said to be *doubly degenerate*, and these are labelled e. Three energy levels with the same energy are said to be *triply degenerate*: these are labelled t. g (and u) refer to the behaviour of an orbital under inversion through a centre of symmetry, and thus are only used for complexes with a centre of symmetry. The d orbitals appear identical after inversion through the centre of symmetry and are therefore labelled g. The orbitals can be further distinguished by subscripts 1 or 2 (such as the 2 in t_{2g}) or by primes ′ or ″.

5.3.1 Strong-field and weak-field complexes

We can determine how electrons occupy the energy levels by:

i determining the oxidation state of the metal ion in the complex

ii calculating the corresponding number of d electrons

iii establishing how these electrons occupy the energy-level diagram (bearing in mind that each energy level can hold a maximum of two electrons of opposite spin).

Firstly, consider a complex of titanium in its +3 oxidation state, which has only one 3d electron (Ti^{3+}, $3d^1$). This electron enters the t_{2g} level (Figure 5.6a). It does not matter whether the electron is in the $3d_{xy}$, $3d_{yz}$ or $3d_{xz}$ orbital because they are degenerate.

For complexes containing metal ions of configuration $3d^2$ (Ti^{2+} and V^{3+}) or $3d^3$ (V^{2+} and Cr^{3+}), the electrons enter the t_{2g} level, but they occupy separate orbitals with parallel spins (Figure 5.6b and c).

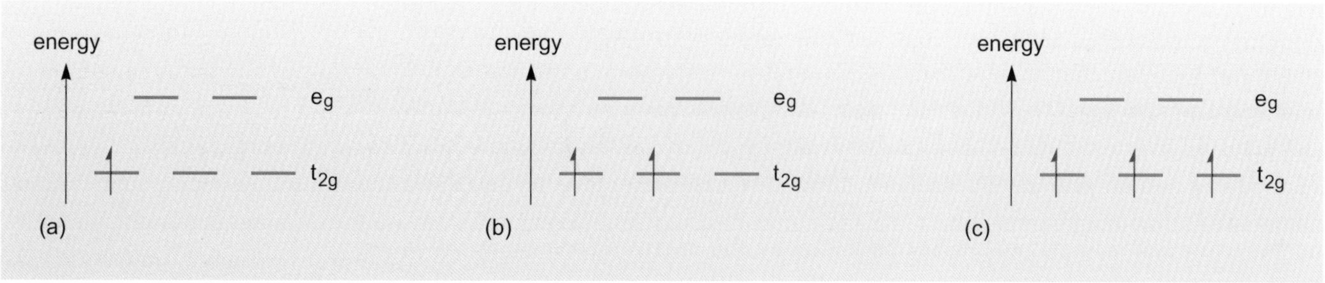

Figure 5.6 Occupation of 3d orbitals for: (a) d^1; (b) d^2; and (c) d^3 ions.

In the $3d^4$ situation, there are two choices. The fourth electron could either enter the t_{2g} level and pair with an existing electron, or it could occupy the e_g level. Which of these possibilities occurs depends on the relative magnitude of two different factors: (i) the crystal-field splitting energy, Δ_o, and (ii) the **pairing energy, *P*.**

Two electrons in one orbital repel each other more than two electrons in different orbitals because, on average, they are closer together. In addition, electrons with paired spins repel each other more than those with parallel spins. Consequently, electron repulsion is minimised if the electrons are in different orbitals with parallel spins. The energy required to force two electrons into the same orbital is the pairing energy, *P*. Thus the following two situations arise.

(a) $\Delta_o < P$: the fourth electron goes into the e_g level, with a spin parallel to those of the t_{2g} electrons. This is known as the **weak-field** or **high-spin** case, and is written $t_{2g}^{\ 3}e_g^{\ 1}$.

(b) $\Delta_o > P$: the energy required for an electron to occupy the upper level, e_g, outweighs the effect of electron–electron repulsion, and the fourth electron goes into a t_{2g} orbital, where it has to be spin paired. This is written $t_{2g}^{4}e_{g}^{0}$, and is known as the **strong-field** or **low-spin** case.

Whether a complex adopts a high-spin or low-spin configuration depends on the metal and the ligand. You will learn more about this later, but it is worth noting here that for complexes of the first-row transition elements, ligands bonding to the metal ion through F or O tend to favour the high-spin configuration.

Now try sketching the orbital energy-level diagrams showing the weak-field and strong-field configurations for d^4–d^7 complexes. These arrangements are illustrated in Figure 5.7.

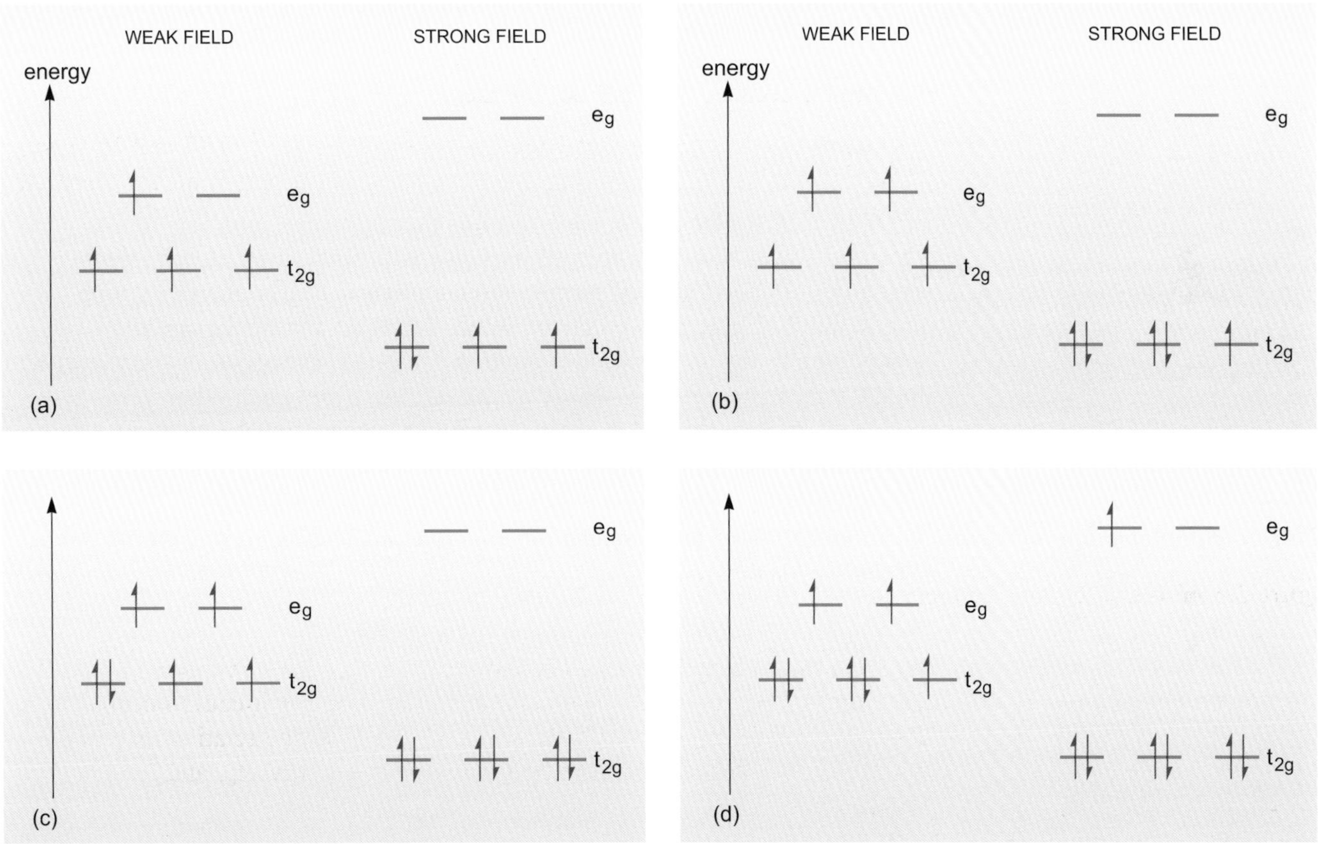

Figure 5.7 Occupation of 3d orbitals for (a) d^4, (b) d^5, (c) d^6 and (d) d^7complexes in weak and strong octahedral crystal fields.

We would not expect to see high-spin and low-spin complexes for d^8 and d^9 complexes in an octahedral crystal field because there is only one possible arrangement of electrons in both cases, the size of Δ_o making no difference to the occupation of the levels. Thus d^8 is $t_{2g}^{6}e_{g}^{2}$ and d^9 is $t_{2g}^{6}e_{g}^{3}$ (see Figure 5.8).

Can the energy-level diagram be used to explain the properties of transition-metal complexes? We shall start by considering ionic radii.

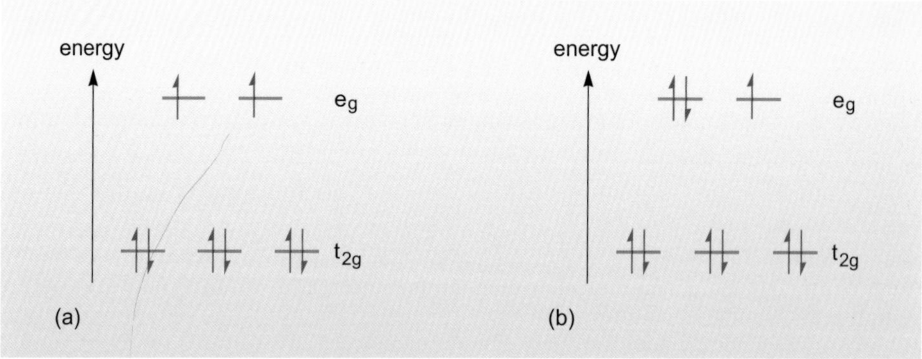

Figure 5.8 Occupation of 3d orbitals for (a) d^8 and (b) d^9 complexes in both weak and strong octahedral crystal fields.

5.3.2 Ionic radii

In Figure 5.9, a plot of the ionic radii of the dipositive ions for the first transition series shows an overall decrease with increasing atomic number, but with a double-bowl shaped profile. As the nuclear charge increases with atomic number, electrons enter the same, 3d, shell and hence are all approximately the same distance from the nucleus. Electrons in the same shell do not screen the positive charge of the nucleus from each other very effectively and so the *net* nuclear charge experienced by the electrons *increases* as the atomic number increases. This increased effective charge pulls the electrons in closer to the nucleus, and hence the ionic radii of the first-row transition elements show an *overall* decrease across the series from about 100 pm to 70 pm.

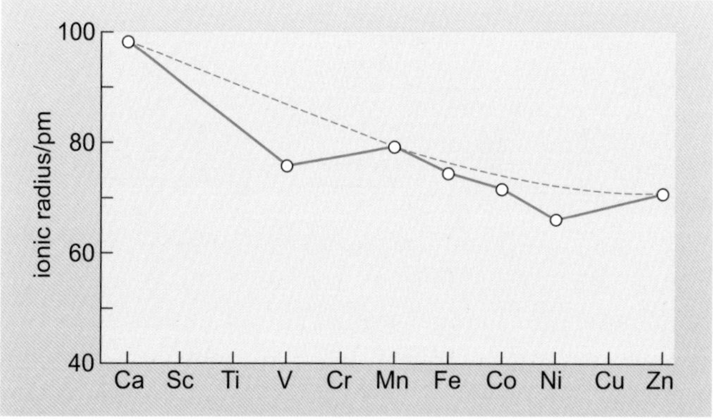

Figure 5.9 Ionic radii of the divalent ions of calcium and the first-row transition metals in the difluorides.

If there were a spherical distribution of electric charge over the ions, we would expect a regular decrease in ionic radii (shown by the dashed red line in Figure 5.9), but the actual values of the radii (the thicker red line) indicate that clearly this is not generally so, except for d^0, d^5 (high-spin) and d^{10}, where there is a regular distribution of the d orbitals. Crystal-field theory explains this. The radii in Figure 5.9 were obtained by measuring metal–fluorine distances in crystalline metal difluorides where each metal ion is surrounded by an octahedral arrangement of fluoride ions. To a first approximation, therefore, we are still dealing with octahedral complexes, and the d-orbital energy-level diagram derived in the previous section (Figure 5.3) applies.

Look first at the point on the graph for V^{2+} in VF_2. The V^{2+} ion has three 3d electrons in the t_{2g} level, and in an octahedral crystal field they occupy the orbitals with parallel spins. The electrons on the transition metal screen the fluoride ions (regarded as point negative charges) from the positive charge of the metal nucleus. Thus the fluoride ions experience a net charge which is less than the charge of the bare nucleus. As the electrons are dividing their time between t_{2g} orbitals that are concentrated between the ligands, this screening is less efficient than if the electrons were in e_g orbitals. Since the fluoride ions are screened less than we would expect if the crystal field were spherical, they experience a greater net charge and move closer to the metal nucleus, hence shortening the metal–fluorine distance. V^{2+}, therefore, has a smaller ionic radius than would be expected for a spherical ion (in fact it has the largest deviation from the spherical ion prediction of all the elements in the first transition series).

■ Use Figure 5.9 to estimate the notional ionic radii of the first two transition metals, the crystalline difluorides of which, ScF_2 and TiF_2, are actually unknown.

□ The argument is the same as for V^{2+} but now for one and for two d electrons in the t_{2g} orbitals, respectively. We expect Sc^{2+} and Ti^{2+} to have smaller ionic radii than if they were spherical ions but with slightly reduced screening effects; the graph predicts about 91 pm and 82 pm, respectively.

Notice that the plot in Figure 5.9 does not include points for Cr and Cu. The reason for this is that chromium(II) fluoride and copper(II) fluoride have distorted octahedral structures. Why this is so will become apparent in Section 5.5; however, we can still use Figure 5.9 to make useful predictions. Cr^{2+} has four d electrons. Fluoride ions are very weak-field ligands, so the complex would be high-spin, with the fourth electron going into the e_g level. The electron density is now partially in an orbital pointing directly at the ligands which would therefore screen the nuclear charge more efficiently; overall the screening would be more than for V^{2+} (d^3) and the Cr^{2+} ion is therefore expected to have a smaller radius than in a spherical environment – but larger than that of V^{2+}.

■ Why is the ionic radius for Mn^{2+} that expected for a spherical ion?

□ Mn^{2+} is a d^5 ion, and in the weak-field (high-spin) fluoride has three electrons in t_{2g} and two in e_g. All five d orbitals are equally occupied, and so the total 3d electron distribution around the Mn^{2+} ion is spherical.

In Fe^{2+}, Co^{2+} and Ni^{2+}, the t_{2g} level is gradually filled, and, like Ti^{2+} and V^{2+}, these ions are smaller than expected. Cu^{2+} has its ninth electron in the e_g

level, but this level is still not full, and therefore the radius is also less than expected for a spherical ion.

Crystal-field theory thus helps us to understand the double-bowl variation of ionic radii across the first transition series. In the next section, we shall see how it can be used to explain the variation of other properties.

5.3.3 Crystal-field stabilisation energy

We now turn our attention to the variation of the **lattice energy** of the first transition series metal chlorides, MCl_2. The lattice energy is defined as the standard enthalpy change $\Delta H_m^{\ominus}$ for the formation of the solid from gaseous ions

$$M^{2+}(g) + 2Cl^{-}(g) = MCl_2(s)$$

These values are plotted in Figure 5.10, which also has a double-bowl shape. This is not surprising as lattice energy depends inversely on ionic radius, which, as you will recall from Figure 5.9, exhibits a similar variation. Again, we can account for this variation with reference to crystal-field theory.

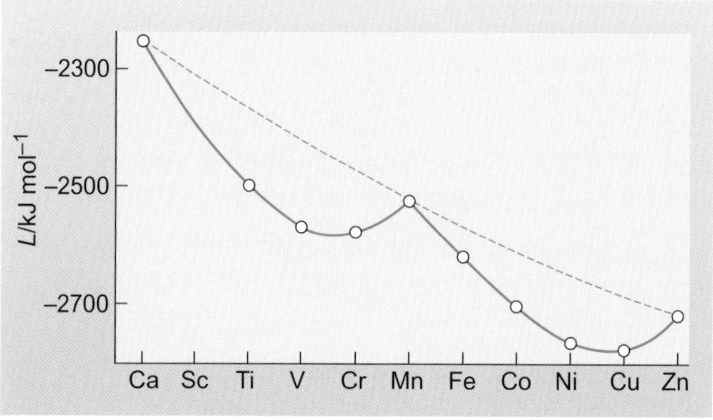

Figure 5.10 Lattice energies of the dichlorides of calcium and the first-row transition metals.

In the dichlorides, the metal ions are in octahedral sites. Considering $TiCl_2$, for example, the ion Ti^{2+} has two d electrons in the t_{2g} orbitals with parallel spins. In Section 5.2.1, we noted that transition-element electrons in octahedral complexes are in a t_{2g} orbital with an energy $\frac{2}{5}\Delta_o$ less than the barycentre, but those in an e_g orbital are increased in energy by $\frac{3}{5}\Delta_o$. The orbital energy for Ti^{2+} in $TiCl_2$ is thus $2 \times \frac{2}{5}\Delta_o$ or $\frac{4}{5}\Delta_o$ below what we would expect for the ion if it were in a (hypothetical) spherical crystal field. This decrease in energy on going from the spherical situation to an octahedral crystal field is called the **crystal-field stabilisation energy, CFSE**.

- ■ What is the CFSE for V^{2+} in an octahedral site?

- □ V^{2+} has three 3d electrons, and so these all go into the t_{2g} orbitals with parallel spins. The CFSE for V^{2+} is thus $3 \times \frac{2}{5}\Delta_o = \frac{6}{5}\Delta_o$. The lattice energy value for VCl_2 in Figure 5.10 is therefore further below the curve for an ion in a spherical environment (the dashed red line) than that for $TiCl_2$, due to the additional contribution from the CFSE.

The magnitude of the CFSE is small (about 10%) compared with the total lattice energy, but nevertheless, CFSE does make its presence felt as we move across a transition series.

For a d^4 ion such as Cr^{2+} in $CrCl_2$, whether it is in a weak or strong field makes a difference.

- ■ What is the CFSE for Cr^{2+} in a *weak* octahedral field?
- □ In the weak-field case, you saw that a $3d^4$ ion has three electrons in t_{2g} and one in e_g ($t_{2g}{}^3e_g{}^1$). This gives a CFSE of $\frac{6}{5}\Delta_o - \frac{3}{5}\Delta_o = \frac{3}{5}\Delta_o$.

In a strong field, all four electrons go into the t_{2g} orbitals, and so the CFSE is $4 \times \frac{2}{5}\Delta_o = \frac{8}{5}\Delta_o$. However, the orbital energy is not the only factor to be taken into account in the strong-field case. The pairing energy acts to *reduce* the CFSE, and so we need to subtract *P for each pair of spins additional to those paired in the free ion*. Therefore, the total CFSE for Cr^{2+} in a strong-field case is $\frac{8}{5}\Delta_o - P$, because there are no paired electrons in the free ion and one pair in the strong-field octahedral environment.

- ■ What are the CFSEs for strong- and weak-field $Mn^{2+}(3d^5)$?
- □ In a strong field, Mn^{2+} has all five electrons in t_{2g}, with two electron pairs, so the CFSE is $(5 \times \frac{2}{5}\Delta_o - 2P) = 2\Delta_o - 2P$. In the weak-field case, Mn^{2+} has the configuration $t_{2g}{}^3e_g{}^2$ where all the spins are parallel. The CFSE is thus zero.

The CFSEs for d^1–d^{10} configurations in weak and strong fields are given in Table 5.2.

Table 5.2 CFSEs for first transition series ions in weak and strong octahedral fields*.

Configuration	CFSE (weak-field)	CFSE (strong-field)	Configuration	CFSE (weak-field)	CFSE (strong-field)
d^1	$\frac{2}{5}\Delta_o$	$\frac{2}{5}\Delta_o$	d^6	$\frac{2}{5}\Delta_o$	$\frac{12}{5}\Delta_o - 2P$
d^2	$\frac{4}{5}\Delta_o$	$\frac{4}{5}\Delta_o$	d^7	$\frac{4}{5}\Delta_o$	$\frac{9}{5}\Delta_o - P$
d^3	$\frac{6}{5}\Delta_o$	$\frac{6}{5}\Delta_o$	d^8	$\frac{6}{5}\Delta_o$	$\frac{6}{5}\Delta_o$
d^4	$\frac{3}{5}\Delta_o$	$\frac{8}{5}\Delta_o - P$	d^9	$\frac{3}{5}\Delta_o$	$\frac{3}{5}\Delta_o$
d^5	0	$2\Delta_o - 2P$	d^{10}	0	0

* Note that the CFSE, as defined, is the decrease in energy on going from a spherical crystal field to an octahedral crystal field. Thus the energy for an electron in a t_{2g} level (for a d^1 complex, for example) is $-\frac{2}{5}\Delta_o$ relative to the barycentre (see Figure 5.3).

■ Account for the CFSE for a d^6 ion in a strong field.

□ In a strong field, a d^6 ion has all six electrons in t_{2g}, with three electron pairs. As the free ion has one electron pair, there are only two additional pairs of spins. Thus the CFSE is $6 \times \frac{2}{5}\Delta_o - 2P = \frac{12}{5}\Delta_o - 2P$.

Like the fluoride ion, the chloride ion produces a weak field, and the lattice energies in Figure 5.10 are all for high-spin complexes. You can see that the deviation from the spherical environment curve (dashed red line) increases from Ti^{2+} to V^{2+}, then decreases at Cr^{2+}. Mn^{2+} lies on the curve, as it has zero CFSE. From Fe^{2+} through to Zn^{2+} (filled d sub-shell) the pattern is repeated, as first the t_{2g} and then the e_g levels are filled.

5.4 Electronic spectra of octahedral complexes

As mentioned previously, one of the most distinctive properties of transition-metal complexes is their wide range of colours. This means that some of the visible wavelengths of the solar spectrum are being absorbed as they pass through the sample, so the visible radiation that emerges is no longer perceived as white light. The colour of the complex is said to be *complementary* to that which is absorbed, and is the colour generated by the receptors in the retina from the wavelengths that are not absorbed. For example, if white light strikes an object, and the green wavelengths are absorbed by it, then the remaining wavelengths reaching the eye are seen as red; green is said to be the complementary colour to red.

■ The complex $[Ti(H_2O)_6]^{3+}$ absorbs yellow–green light. What colour does it appear? (Think where in the spectrum the green and yellow wavelengths occur and therefore what will be left if they are removed.)

□ The green and yellow wavelengths are in the middle of the visible spectrum, so the red and blue ends remain, generating a violet colour (Figure 5.11).

Next we need to ask, what are the origins of these light-absorbing processes in metal complexes?

5.4.1 d–d transitions

As we have seen, in octahedral complexes the orbitals are split into a lower-energy t_{2g} set and higher-energy e_g set. **Electronic transitions** can occur between these partially filled d orbitals; an electron in the lower set can be given sufficient energy to make it jump up to the higher energy levels. These are called d↔d or **d–d transitions** (pronounced 'dee dee') and are one source of colour for transition-metal compounds. The frequency of the light absorbed, and thus the colour observed, is a measure of the crystal-field splitting, Δ_o, because the change of energy, ΔE, and frequency of light, ν, are related by the following equation, where h is Planck's constant

$$\Delta E = h\nu \qquad (5.1)$$

Figure 5.11 shows the electronic spectrum of $[Ti(H_2O)_6]^{3+}$, a d^1 ion, and so the simplest example we can take. The horizontal scale is wavenumber, for which the common symbol is σ, and which is measured in cm^{-1}.

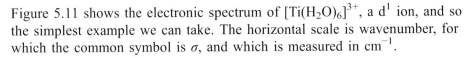

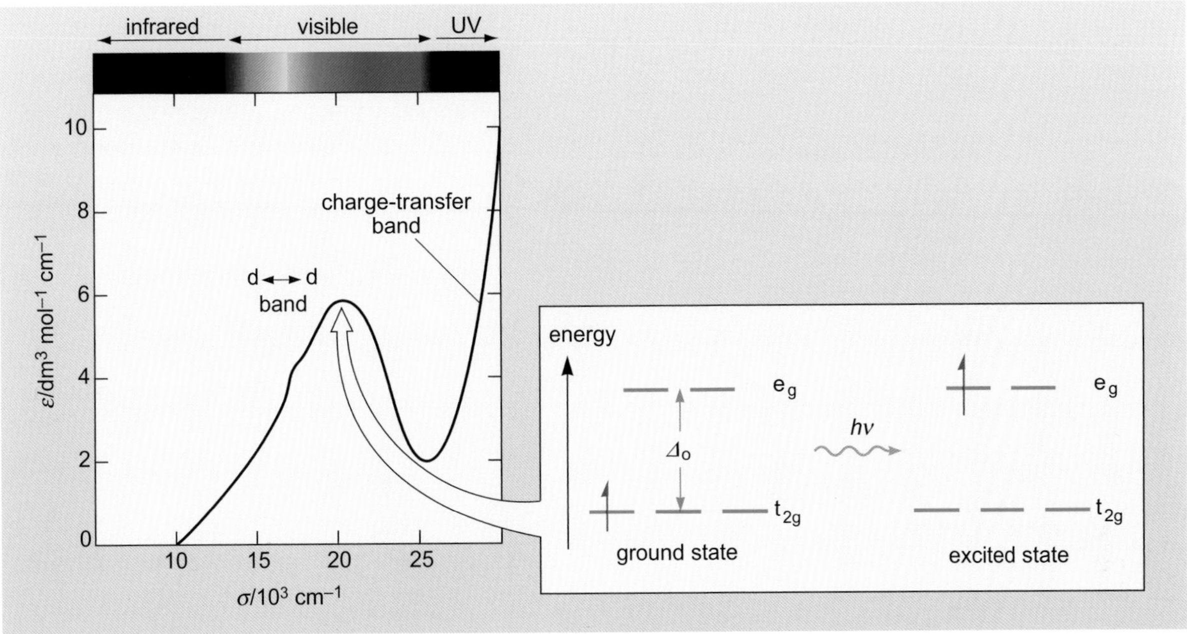

Figure 5.11 Electronic absorption spectrum of $[Ti(H_2O)_6]^{3+}$; the inset shows the transition of an electron from the t_{2g} level to the e_g level, which gives rise to the d–d band.

The peak in Figure 5.11 is due to an electron jumping (or being excited) from t_{2g} to e_g in the $[Ti(H_2O)_6]^{3+}$ ion, and may be represented by the energy-level diagram shown in the inset. (Ignore for the moment the intense peak labelled 'charge-transfer band', we will consider this later in Chapter 6.) The d–d transition of the electron in t_{2g} being excited up to the e_g level confers the violet colour on the complex. Measuring the wavenumber of maximum absorption for a complex with a single 3d electron should therefore give us a value of Δ_o directly.

Note that the molar absorption coefficient, ε, for this band is 6 $dm^3\ mol^{-1}\ cm^{-1}$, which represents a very low value, characteristic of d–d spectra. This spectrum is also very broad, another feature frequently observed in d–d spectra: this width is due to metal–ligand vibrations – analogous to the rotational fine structure of vibrational spectra.

The absorbance, A, the quantity measured, is given by $A = \varepsilon cl$ where c is the concentration of the solution ($mol\ dm^{-3}$), l is the path length of the cell (cm) and ε is the molar adsorption coefficient ($dm^3\ mol^{-1}\ cm^{-1}$).

5.4.2 The spectrochemical series

We have already seen that transition-metal complexes exhibit a variety of colours, which are dictated by the value of Δ_o. It would therefore be of interest to examine more closely the factors that affect the size of crystal-field splitting.

Figure 5.12 shows schematic spectra for titanium(III) complexes involving different ligands; notice that the peaks corresponding to the d–d transitions clearly occur at different wavenumbers and therefore at different energies. Weak-field ligands have a small Δ_o and strong-field ligands have a large Δ_o.

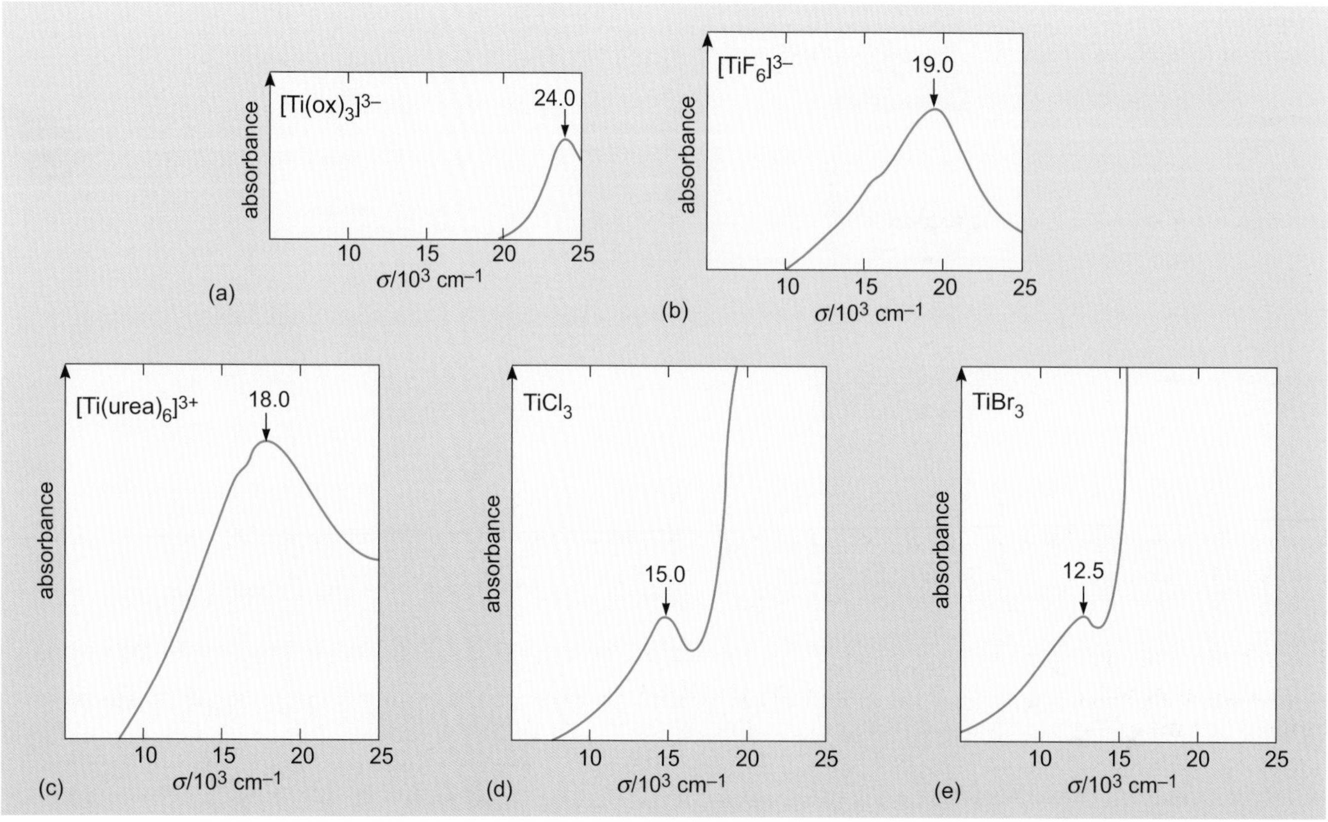

Figure 5.12 Schematic electronic absorption spectra of: (a) $[Ti(ox)_3]^{3-}$; (b) $[TiF_6]^{3-}$; (c) $[Ti[(urea)_6]^{3+}$; (d) $TiCl_3$; and (e) $TiBr_3$.

■ With reference to Figures 5.11 and 5.12, arrange the ligands H_2O, F^-, urea, Cl^-, Br^- and oxalate (ox, $(COO^-)_2$) in order, from the weakest-field to the strongest-field ligand.

□ The larger the value of Δ_o, the more energy is absorbed in the transition from t_{2g} to e_g, and the higher the frequency/wavenumber of the spectral band. The order here is thus:

$$Br^- < Cl^- < \text{urea} < F^- < H_2O < \text{ox}$$

The strong-field ligands towards the end of the series are π-acceptors.

A series such as this, which arranges ligands in order of their Δ_o, is known as a **spectrochemical series**. By studying a large number of complexes, it is possible to compile a series that is valid for complexes of these ligands with most transition metals. A more extensive series is given below with increasing Δ_o from left to right (weak-field to strong-field):

(weak-field) $I^- < Br^- < SCN^- < Cl^- < S^{2-} < F^- <$ urea $\approx OH^- <$ ox $\approx O^{2-} < H_2O < NCS^- < \text{edta}^{4-} < NH_3 <$ en $<$ bipy $\approx$ phen $\approx NO_2^- < PR_3 < CN^- < CO$ (strong-field)

Note that SCN⁻ represents a thiocyanate anion bound to a metal through S, and NCS⁻, isothiocyanate, the same anion bound through N. In PR_3, R is used to represent a saturated alkyl group. Although you have met them in Chapter 3 (Table 3.3), the ligands $edta^{4-}$, en, bipy and phen are shown again in structures **5.1** to **5.4**; the coordinating atoms are shown in colour.

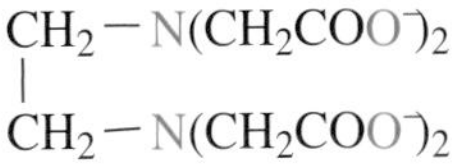

$$CH_2-N(CH_2COO^-)_2$$
$$CH_2-N(CH_2COO^-)_2$$

5.1

$$H_2N \qquad NH_2$$
$$CH_2-CH_2$$

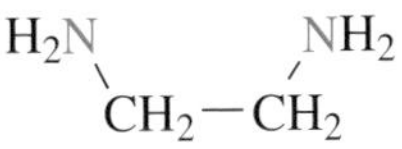

5.2

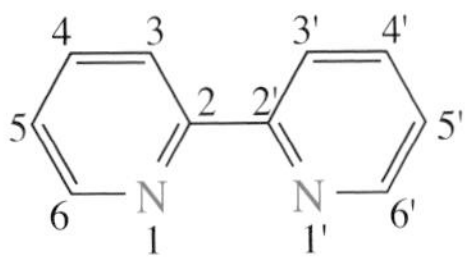

5.3

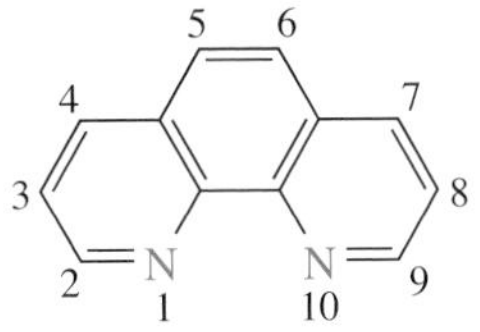

5.4

You can roughly link the series with the position of the coordinating atom in the Periodic Table: crystal-field strength decreases from Group 14 to Group 17.

The influence of the ligand on the colour of complexes can be seen in a series of octahedral nickel(II) complexes. Consider the $[Ni(H_2O)_6]^{2+}$ complex which forms when nickel(II) chloride is dissolved in water; when the bidentate ligand 1,2-diaminoethane (en) is progressively added in the molar ratios en:Ni of 1:1, 2:1 and 3:1 (where en progressively replaces two H_2O ligands) the following colour changes occur:

$$[Ni(H_2O)_6]^{2+}(aq) + en(aq) = [Ni(H_2O)_4en]^{2+}(aq) + 2H_2O(l) \qquad (5.2)$$
$$\text{green} \qquad\qquad\qquad \text{pale blue}$$

$$[Ni(H_2O)_4en]^{2+}(aq) + en(aq) = [Ni(H_2O)_2en_2]^{2+}(aq) + 2H_2O(l) \qquad (5.3)$$
$$\text{blue/purple}$$

$$[Ni(H_2O)_2en_2]^{2+}(aq) + en(aq) = [Nien_3]^{2+}(aq) + 2H_2O(l) \qquad (5.4)$$
$$\text{violet}$$

This sequence is shown in Figure 5.13.

It is interesting to stop and think about how well this spectrochemical series ties in with the purely electrostatic (repulsive) treatment, which forms the basis of crystal-field theory. By this approach we would expect more highly charged ligands to produce stronger fields, and consequently have larger values of Δ_0. However, the charged halides produce very weak fields, whereas the strongest field is produced by CO, a neutral molecule. Even water produces a stronger field than its charged relatives OH^- and O^{2-}. Thus, crystal-field theory gives no indication as to why ligands occupy the positions they do in the series. To account for this we need to consider covalent bonding in the metal–ligand interaction, and, as you will see in Chapter 6, that is where molecular orbital theory will fit in.

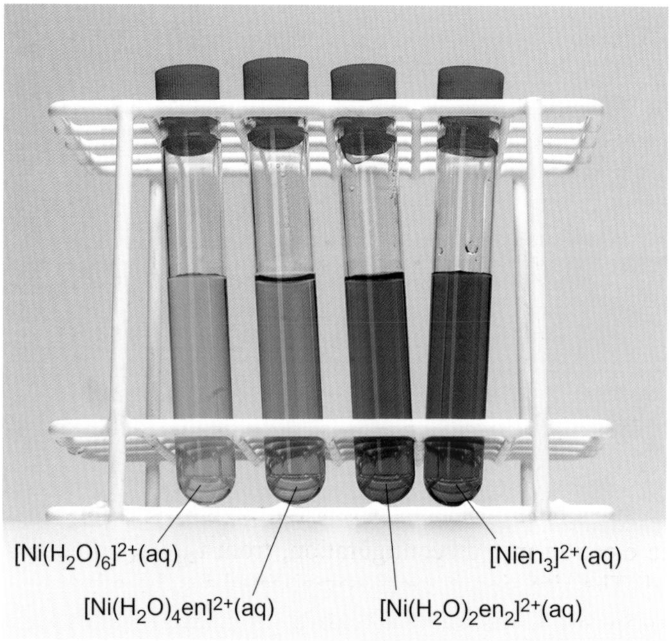

Figure 5.13 Aqueous solutions of complexes of nickel(II) with an increasing number of 1,2-diaminoethane ligands.

Empirically, the size of Δ_o is found to depend not only on the type of ligand, but also on contributions from the central metal:

(a) For a given ligand and a given metal, crystal-field splitting increases with an increase in metal oxidation state; for example, Δ_o is 9400 cm^{-1} for $[Fe(H_2O)_6]^{2+}$, but 13 700 cm^{-1} for $[Fe(H_2O)_6]^{3+}$. On simple electrostatic grounds, we can argue that the increase in the positive charge on the metal draws the ligands closer, which results in greater repulsion between the metal d electrons and the ligand point charges, hence causing an increase in Δ_o.

(b) For a given ligand, crystal-field splitting increases on descending a Group in the Periodic Table (see the data in Table 5.3 for Co, Rh and Ir). In this case, we could conjecture that, as the size of the d orbital increases on descending the Group (Co 3d, Rh 4d, Ir 5d), the d electrons are progressively closer to the ligands, and so Δ_o increases.

Table 5.3 Crystal-field splitting for ammino complexes of the Co, Rh and Ir Group.

Complex	Δ_o/cm^{-1}
$[Co(NH_3)_6]^{3+}$	23 000
$[Rh(NH_3)_6]^{3+}$	34 000
$[Ir(NH_3)_6]^{3+}$	41 000

■ The enzyme ribonucleotide reductase catalyses the reduction of ribonucleotides to deoxyribonucleotides. A spectroscopic study of this process (Mitić et al., 2007) identified an intermediate with Fe in oxidation state +4. The electronic spectrum of this intermediate showed d–d bands in the region 17 360–24 320 cm^{-1}. The surrounding ligands coordinate through O. Are the positions of these bands consistent with the generalisation of the variation of Δ_o with oxidation state?

□ Yes. The wavenumber of d–d transitions depends on the crystal-field splitting and hence on Δ_o. This is expected to increase with oxidation state. You have seen that Δ_o is 9400 cm^{-1} for $[Fe(H_2O)_6]^{2+}$, but 13 700 cm^{-1} for $[Fe(H_2O)_6]^{3+}$. As the intermediate also has O-donating ligands, a further increase is expected for an Fe(IV) complex.

5.4.3 Spectra of complexes with more than one d electron on the metal ion

Transitions of the single electron in a d^1 configuration, from $t_{2g} \rightarrow e_g$, produce a single absorption band. The situation is more complicated for the other configurations, as shown by the spectrum of a d^5 manganese(II) complex in Figure 5.14.

The value of Δ_o for $[Mn(H_2O)_6]^{2+}$ cannot be obtained simply from this spectrum as it contains several d–d bands. This is because when there is more than one d electron, we need to take into account not only the orbital energy change, but also the change in repulsion energy and the change in spin state between the d electrons, when a transition occurs.

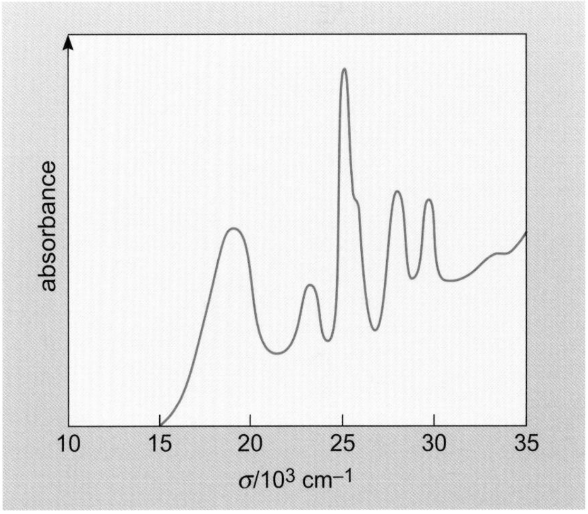

Figure 5.14 Electronic absorption spectrum of $[Mn(H_2O)_6]^{2+}$.

The different way electrons can occupy the orbitals in a particular configuration is referred to as a **state**. The different possible states are defined by the total orbital quantum number, L, and total spin quantum number, S, for the ion. The same letters are used as for the orbital quantum number, l, i.e. s, p, d, f, but in upper case, so for example the state where $L = 3$ is called an F state.

It is not a simple matter to identify the transitions giving rise to d–d bands for complexes with more than one d electron, but a consideration of the d^2 case will give you an idea of how the electron repulsion terms affect the spectra. A d^2 ion has bands corresponding to one of its d electrons undergoing a transition from t_{2g} to e_g. Supposing that the electron remaining in the t_{2g} level is in a d_{xy} orbital, then the electron promoted to the e_g level can be either in $d_{x^2-y^2}$ or d_{z^2}.

■ Would the repulsion between an electron in d_{xy} and one in $d_{x^2-y^2}$ be the same as that between one in d_{xy} and one in d_{z^2}?

□ No: electrons in d_{xy} and $d_{x^2-y^2}$ are closer to each other on average than ones in d_{xy} and d_{z^2} (see Figure 5.1). Therefore, we would expect a greater repulsion between electrons in d_{xy} and $d_{x^2-y^2}$.

A different amount of energy would therefore be needed to promote an electron from t_{2g} to $d_{x^2-y^2}$ than that for a transition from t_{2g} to d_{z^2}, and so we would expect to see two bands instead of one for this particular $t_{2g} \rightarrow e_g$ transition.

- Can we be sure that the electron remaining in t_{2g} is in d_{xy}?

- No: it could be in d_{yz} or d_{xz} or d_{xy}.

Similar arguments can be applied if the electron is in d_{xz} or d_{yz}. Whichever t_{2g} orbital the electron occupies, there is still a difference in repulsion energy depending on which orbital the e_g electron occupies. For d^2 there are five possible states dependent on which orbitals the electrons occupy, and many more when electron spin is also taken into account. In the d^2 $[V(H_2O)_6]^{3+}$ ion, because of selection rules (Section 5.4.4), only three bands are possible: two are observed at 17 800 cm^{-1} and 25 700 cm^{-1}, corresponding to a transition of an electron from t_{2g} to e_g, and the third is obscured by an intense charge-transfer band (see Chapter 6). Note that none corresponds exactly to Δ_o, which for this complex is 19 200 cm^{-1}.

With more than two electrons, the spectra become even more complicated as shown, for example, in Figure 5.14. Manganese(II) complexes have five d electrons, and even for a simple complex such as $[Mn(H_2O)_6]^{2+}$, there are seven d–d bands, six of which are apparent in the spectrum.

5.4.4 Selection rules

As we have seen, d–d bands in the visible spectrum are very weak; that is, they have small molar absorption coefficients, ε. Just as in rotational and vibrational spectroscopy, there are selection rules which determine if a particular transition is allowed. In essence, selection rules dictate the probability of a transition occurring, and even though particular d–d bands are formally forbidden, there are mechanisms that can account for the occasional photon being absorbed and weak colour being observed.

The **Laporte selection rule** requires that during an electronic transition the orbital quantum number, l, can only change by ±1. In other words, we can have transitions from an s orbital to a p orbital, s→p, p→d, etc., but not s→d where the change is +2, or indeed, the d↔d bands we have just been discussing because the change is 0; the latter two transitions are said to be **forbidden transitions**. When we study molecular orbitals for complexes in Chapter 6, we shall see that the orbitals are not pure d atomic orbitals and this leads to some relaxation of this rule. A more general rule states that if a complex has a centre of symmetry, a transition from a g orbital to another g orbital, or from a u orbital to another u orbital, is forbidden. Thus, in complexes with octahedral symmetry the transition from t_{2g} to e_g is forbidden, because both sets of orbitals have the same symmetry with respect to inversion through the centre of symmetry. However, there is a relaxation of this rule due to interaction with vibrational motion of the ligands in the metal ion coordination sphere.

Electronic transitions must also obey the **spin selection rule**, which states that an electron cannot change its spin while undergoing a transition to another level. Considering a d–d transition of a high-spin d^5 ion, it is clear that if an electron is to be excited from t_{2g} to e_g, it must change its spin as there is no room in the upper level for an electron with the same spin (Figure 5.15). Such a transition is said to be spin-forbidden.

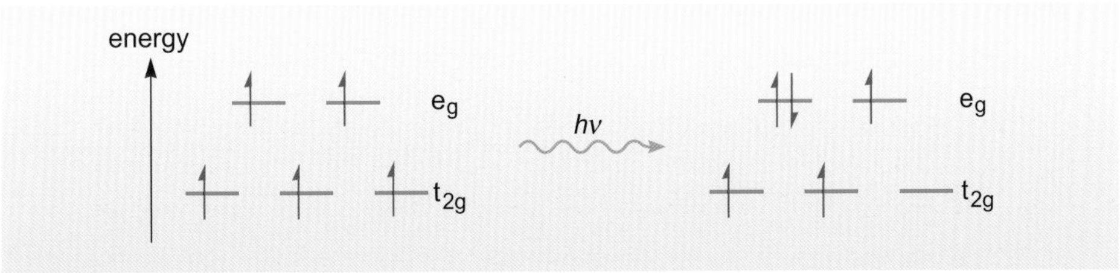

Figure 5.15 A spin-forbidden d–d transition for a d^5 ion in a weak-field octahedral complex.

Not all d–d transitions are of course spin-forbidden. For example, in a d^1 ion, an electron can be excited from t_{2g} to e_g without changing its spin (Figure 5.11).

To see what this means in practice, think of an equimolar solution of the complexes of d^7 $[Co(H_2O)_6]^{2+}$ and d^5 $[Mn(H_2O)_6]^{2+}$ (i.e. equal molar concentrations of these two complexes). Although the former is not strongly coloured, the latter is virtually colourless and, as a powdered solid, appears white. For a high-spin d^7 ion like Co^{2+}, an electron can be promoted from a t_{2g} orbital to an e_g orbital without breaking the spin selection rule (it still breaks the Laporte d→d rule *and* the centre of symmetry g→g rule). However, Mn^{2+} is a d^5 ion, and, as shown in Figure 5.15, promotion is impossible in a high-spin environment without reversing the electron spin in the excited state, so in this case d–d transitions are forbidden by the Laporte d→d rule, the g→g rule *and* the spin selection rule.

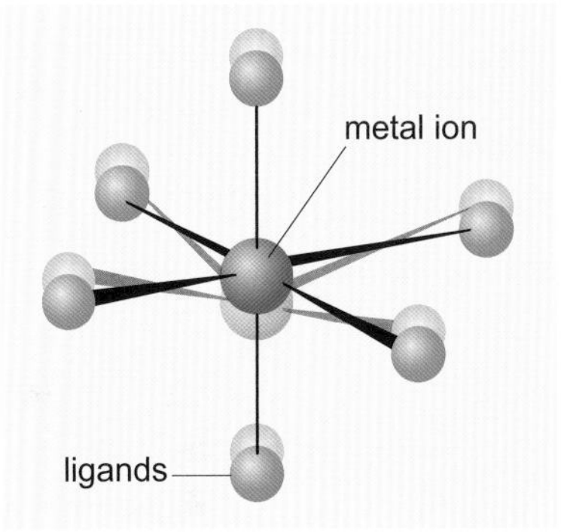

Figure 5.16 An asymmetric vibration in an octahedral complex, leading to the removal of the centre of symmetry. The positions of the atoms at two stages of the asymmetric vibration are superimposed; dark colours are used for the first positions of the atoms, and light colours for the later positions. In the first position there is a centre of symmetry, but this is not present in the second.

As the d–d transitions in a complex such as $[Mn(H_2O)_6]^{2+}$ are forbidden both by the Laporte and spin selection rules, we might query why we see them at all. The answer is that even transitions such as these become partly allowed if we take vibrations into account. For example, certain molecular vibrations may transiently remove the centre of symmetry (Figure 5.16). Hence an electronic transition becomes partially allowed. This is known as **vibronic (vibrational–electronic) coupling**.

Box 5.2 Ruby

The beautiful red colour of ruby (Figure 5.17) is due to a d–d transition of Cr^{3+} impurities in corundum (one form of Al_2O_3), where they occupy distorted octahedral sites. Cr^{3+} is a d^3 ion and thus we have to take electron repulsion into account as you saw for d^2. In this case there are three spin-allowed transitions, two of which occur in the blue/green region of the visible part of the electromagnetic spectrum.

Figure 5.17 A ruby; this gemstone was found in marble from Mogok, Burma; crystal length 12 mm.

■ What colour will the crystal appear if it absorbs in the blue/green region of the visible spectrum?

□ The crystal will, of course, appear red.

When electrons are excited they can return to the ground state by emitting radiation. In the case of ruby, when the electrons are excited from t_{2g} to e_g they can also undergo a radiationless transition in which the energy is transferred to lattice vibrations rather than being emitted as radiation. After this transition, the Cr^{3+} ion is in a second excited state with the electronic configuration ${t_{2g}}^3$ but with two electrons paired.

■ Why is a return to the ground state now less likely?

□ The transition is not only forbidden by the Laporte selection rule, it is now forbidden by the spin selection rule.

This process is the basis of the use of ruby in lasers. Strong irradiation by visible light can lead to a build up of population in the ${t_{2g}}^3$ state with two paired spins so that eventually there will be more ions in this state than in the ground state, a situation known as **population inversion**. When a photon is emitted from one ion, it triggers emission from other ions and an intense coherent beam of light can be produced.

d–d transitions are just one source of colour in transition-metal complexes. In some cases the colour is due to electronic transitions between the orbitals of the ligands. This is so in many biological molecules, for example chlorophyll. The origin of some of the deepest colours found is, however, a transition between an orbital on the ligand and one on the metal. Such charge-transfer transitions will be introduced in Activity 5.1 and described later in Chapter 6 as they require molecular orbital theory to explain them.

Activity 5.1

You should now work through Section 3 of the multimedia activity entitled *Chemistry of the transition elements* in 'Introducing the transition elements (d Block)':

Section 3: Colour and transition metal compounds

This section introduces the origin of colour in transition metal complexes and briefly looks at both d–d transitions (which were covered in Section 5.4.1) and charge-transfer transitions which will be studied in detail in Chapter 6.

5.5 Distorted octahedral, square pyramidal and square-planar complexes

So far, we have only dealt with regular octahedral complexes, in which the metal ion is surrounded by six equidistant, identical ligand atoms. Although convenient for teaching the principles of crystal-field theory, such complexes are more the exception than the rule. Complexes with other geometries, such as tetrahedral or square-planar, are common. Even nominally 'octahedral' complexes may contain two (or even more) different ligands, for example $[TiCl_2(H_2O)_4]^+$, which means that, although they may be based on an octahedral framework, they do not have true octahedral symmetry.

5.5.1 Distorted octahedral complexes (with 2 long and 4 short bonds)

Firstly, consider what happens when an octahedral complex is distorted by gradually lengthening the bonds of two ligands *trans* to each other, and slightly shortening the metal–ligand distances of the other four. Suppose that the longer bonds are on the z-axis; as the bonds lengthen, the repulsion between the electrons in the metal $3d_{z^2}$ orbital and the ligands *decreases*, which results in the energy of the $3d_{z^2}$ orbital being *lowered*.

An electron in $3d_{x^2-y^2}$ is repelled by the ligands on the x- and the y-axis and, as these bonds shorten, the energy of the $3d_{x^2-y^2}$ orbital is *raised*. Thus, the $3d_{z^2}$ and $3d_{x^2-y^2}$, which made up the e_g level in the octahedral complex, are now *no longer at the same energy*; that is, their degeneracy has been lifted, and the $3d_{z^2}$ is at lower energy than the $3d_{x^2-y^2}$. The original e_g level splits into two as shown in the middle column of Figure 5.18.

What about the other three d orbitals? The $3d_{xy}$ (like $3d_{x^2-y^2}$) is still repelled by the now closer ligands on the x- and y-axes, but $3d_{xz}$ and $3d_{yz}$ are lowered in energy because the electrons occupying them experience less repulsion from the more-distant ligands on the z-axis. The three orbitals which made up the t_{2g} level in the octahedral situation have now split into two levels. Note that because the t_{2g} orbitals do not point directly at the ligands, the splitting of the t_{2g} level is much less than that of the e_g level.

5.5.2 Square-pyramidal complexes

Removing one ligand from along the z-axis to give a **square-pyramidal complex** results in the same splitting pattern. Simple transition-metal square pyramidal complexes are not very common. One example is $Fe(acac)_2Cl$, **5.5**.

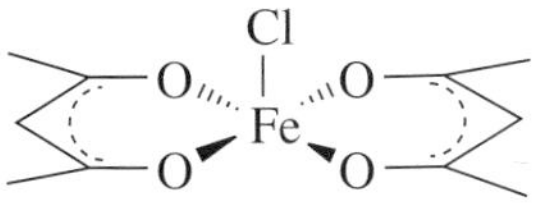

5.5

However, square pyramidal complexes are commonly found in biological systems, the vacant site being reversibly occupied during a reaction or by a molecule that is being transported. For example, deoxyhaemoglobin has a square pyramidal arrangement around an iron atom. When oxygen adds to the protein to form oxyhaemoglobin, it occupies the vacant site to form a near-octahedral complex.

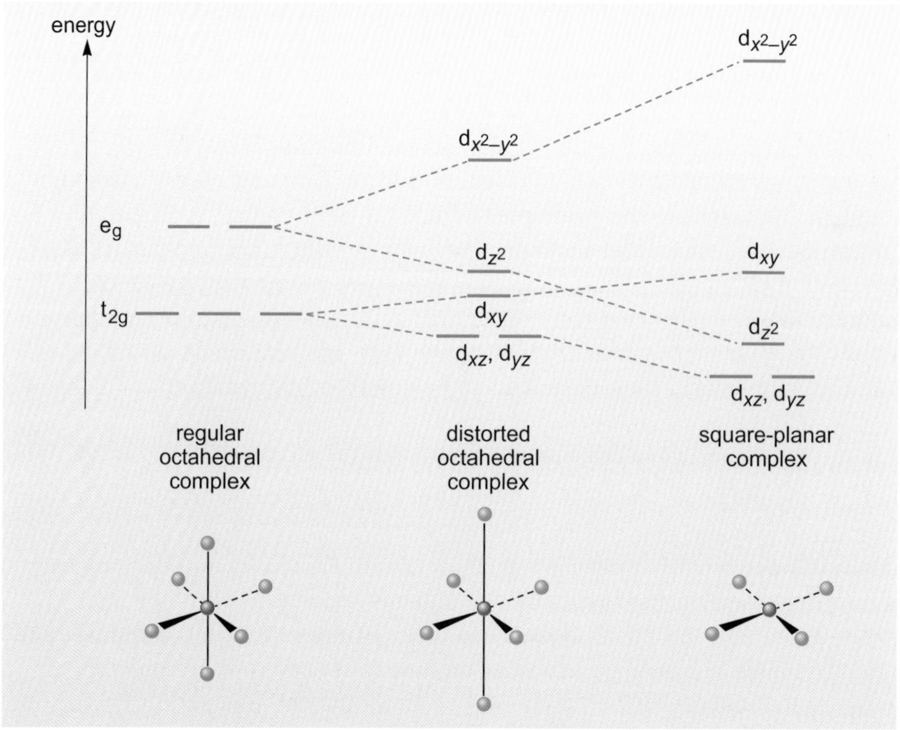

Figure 5.18 Splitting of the 3d levels for a regular octahedral complex, a distorted octahedral complex (with two *trans* bonds longer than the other four) and a square-planar complex.

5.5.3 Square-planar complexes

As the ligands in an octahedral complex move even further out, the splitting of the levels increases with the effect more marked for the d_{z^2} and $d_{x^2-y^2}$ orbitals because they are directed towards the ligands. If the *trans* ligands are removed altogether, the complex becomes square-planar in shape. The above arguments become even more extreme, as illustrated in the right-hand column of Figure 5.18, and the d_{z^2} level may fall below that of d_{xy}.

Square-planar complexes are particularly common at the end of the transition-metal series, especially for ions with d^8 and d^9 configurations. Nickel, palladium and platinum, for example, form many square-planar complexes. One notable example is *cis*-$[PtCl_2(NH_3)_2)]$, commonly known as *cis*-platin (*cis*-diamminodichloroplatinum(II); **5.6**), a potent anticancer drug, which is very effective against testicular and ovarian cancers in particular.

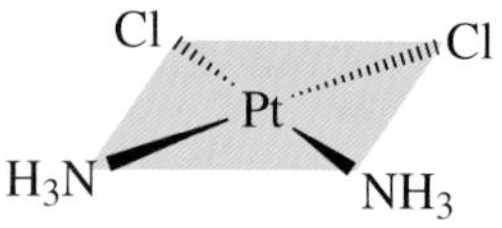

5.6

Referring to the general energy-level diagram in Figure 5.18, you can see why this geometry is favoured by d^8 and d^9 ions. Platinum(II) has eight d electrons, with the configuration $t_{2g}{}^6e_g{}^2$ in a regular octahedral complex. In a square-planar complex with strong-field ligands, the gap between $d_{x^2-y^2}$ and the other d levels is large, and eight electrons fill the d_{xy}, d_{xz}, d_{yz} and d_{z^2} levels, leaving the $d_{x^2-y^2}$ level empty. This means that the highest-occupied energy level is lower in energy than it would be in an octahedral complex. Thus, there is an energy advantage in forming a square-planar complex. Of course, we have to balance this against other factors, such as the pairing of the two electrons in d_{xy} and the different number of metal–ligand bonds in the two geometries. Thus, we would expect square-planar complexes with strong-field ligands where the gain in orbital energy is sufficient to offset these other terms, and in particular for second- and third-row transition elements such as palladium and platinum (as you have seen in Section 5.3, Δ_o is larger for these than for first-row transition elements).

Nickel(II), which has the configuration $3d^8$, forms square-planar complexes with strong-field ligands such as cyanide (CN^-), but four-coordinate complexes of nickel(II) with weak-field ligands are tetrahedral (for example, $[NiCl_4]^{2-}$). Nickel(II) may also form octahedral complexes with weak-field ligands such as halide ions and ligands coordinating through N and O, for example $[NiF_6]^{4-}$ and $[Ni(NH_3)_6]^{2+}$.

5.5.4 Distorted octahedral complexes (with 2 short and 4 long bonds)

Now consider the opposite situation in which the two *trans* ligands move *closer* to the metal ion, and the four in the *xy*-plane move further out. In this case, the *trans* ligands repel an electron in the $3d_{z^2}$ orbital more than they would in a regular octahedral complex.

■ What effect does this have on the $3d_{z^2}$ orbital energy?

□ The $3d_{z^2}$ orbital is even higher in energy than in the regular octahedral complex.

As with the previous example, the e_g level splits into two, but this time the higher level is d_{z^2} and the lower level is $d_{x^2-y^2}$. The t_{2g} level also splits into the higher level (d_{xz}, d_{yz}) and the lower level (d_{xy}). The orbital energy-level diagram for this type of complex is shown in Figure 5.19.

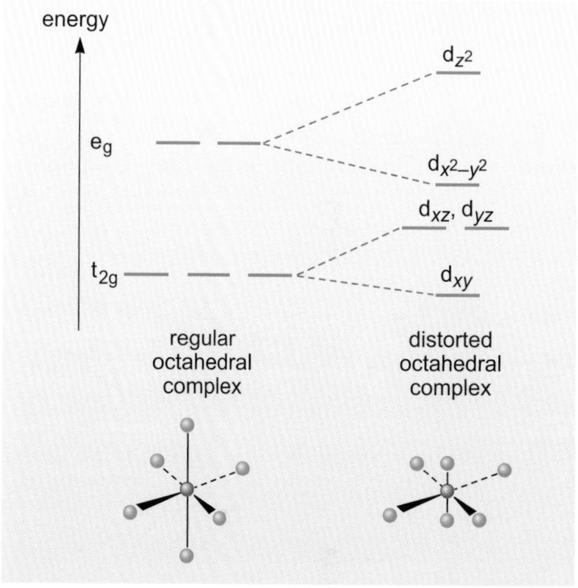

Figure 5.19 Splitting of the 3d levels for a regular octahedral complex and a distorted octahedral complex in which two *trans* ligands on the *z*-axis are closer to the metal ion than the four in the *xy*-plane.

5.5.5 The Jahn–Teller theorem

Typical compounds that adopt distorted octahedral geometries are the halides of copper(II) (Figure 5.20) and chromium(II), which in the solid state have distorted octahedral geometries where the metal ion is surrounded by two elongated bonds opposite each other (*trans*), and four shorter bonds in the *xy*-plane. These complexes are said to exhibit **Jahn–Teller distortion**.

The **Jahn–Teller theorem** says that a non-linear molecule is unstable in a degenerate state and distorts to remove the degeneracy.

We explained in Section 5.4.3 that the different ways electrons can occupy the orbitals of a particular electronic configuration are referred to as *states*. To understand this better, we can illustrate the situation in a high-spin (weak-field) complex of d^4 chromium(II), such as $[Cr(H_2O)_6]^{2+}$.

In a weak-field octahedral complex, these four electrons go into the three t_{2g} orbitals and one e_g orbital with parallel spins. Now, the t_{2g} electrons can be

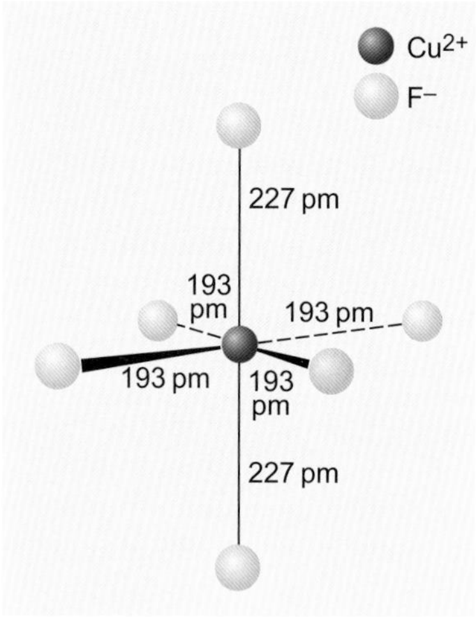

Figure 5.20 An example of Jahn–Teller distortion in copper(II) fluoride.

arranged only in one way: one of them has to be in d_{xy}, one in d_{xz} and one in d_{yz}. However, the e_g electron can be in either d_{z^2} or $d_{x^2-y^2}$. In a regular octahedral complex, the energy of the complex would be the same for either of these alternatives, and so the complex would be in either of two degenerate states (Figure 5.21a). So, complexes of d^4 chromium(II), according to the Jahn–Teller theorem, distort to remove this degeneracy, often by having two ligands further away than the other four (Figure 5.18). When an octahedral complex is distorted in this way, the e_g level splits into two levels of different energy and the electron occupies the lower level removing the degeneracy.

To take another example, d^9 complexes, such as those of copper(II), without distortion would also have degenerate states because, as in the d^4 case, there is a choice of e_g orbital (Figure 5.21b).

■ What other examples will have degenerate states because of a choice of e_g orbital?

□ Low-spin (strong-field) octahedral d^7 complexes would also be degenerate.

d^1 and d^2 complexes (for example, those of titanium(III) and vanadium(III)) would have degenerate states if there were no distortion because of the choice of t_{2g} orbitals to occupy. d^6 and d^7 complexes (for example, high-spin iron(II) and cobalt(II)) also have a choice of t_{2g} orbitals available.

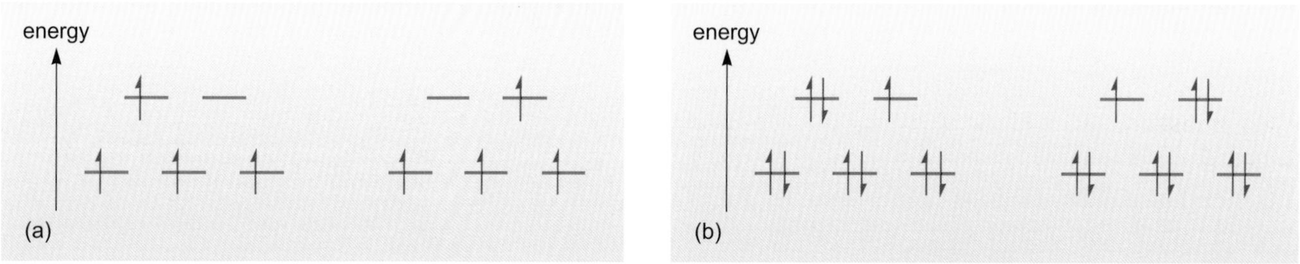

Figure 5.21 The two degenerate states of (a) a d^4 configuration and (b) a d^9 configuration.

How do we explain the Jahn–Teller theorem and why these distortions take place? We saw earlier that as a distorted complex is formed by moving four ligands nearer to the metal and two ligands further away from the metal, *if the total electrostatic interaction between ligands and metal ions remains the same*, the d_{z^2} level must be lower in energy, and the $d_{x^2-y^2}$ higher in energy, than the e_g level in the regular octahedral complex (Figure 5.18).

■ In a high-spin d^4 complex, which levels do the electrons occupy if the complex is Jahn–Teller distorted as we have indicated?

□ The electrons go into the d_{xz}, d_{yz} and d_{xy} from the octahedral t_{2g}, and into d_{z^2} from the octahedral e_g (Figure 5.22).

As d_{z^2} is lower in energy than the regular octahedral e_g level, *the distorted complex is more favourable energetically* than a regular octahedral complex: that is, the highest-energy electron is in a lower energy state than if it were in an undistorted octahedral complex.

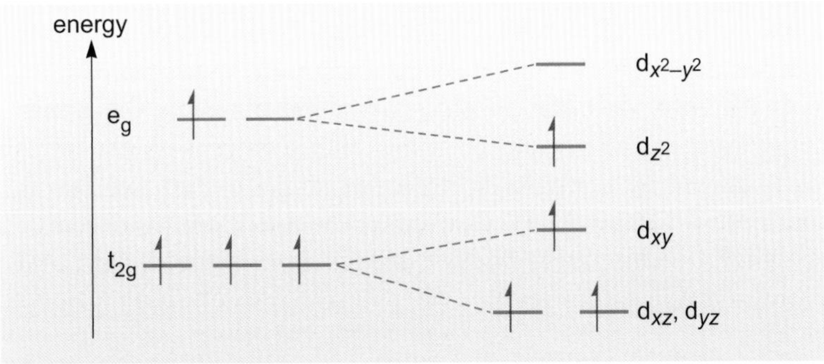

Figure 5.22 Occupation of the t_{2g} and e_g levels for a d^4 Jahn–Teller-distorted octahedral complex.

This explanation holds for d^9 copper(II) complexes as well. For copper(II), two electrons go into the d_{z^2} and one into $d_{x^2-y^2}$, compared with three in e_g for the regular octahedral complex. Complexes of chromium(II) and copper(II) thus distort because there is an *overall reduction* in orbital energy to be achieved on doing so.

The distortions from regular octahedral geometry observed in d^1, d^2, d^4 (low-spin), d^5 (low-spin), d^6 (high-spin) and d^7 (high-spin) complexes are relatively small in comparison with d^4 (high-spin), d^7 (low-spin) and d^9 complexes. This is because t_{2g} orbitals lie between the ligands, so the splitting of the t_{2g} orbitals in the distorted state is much smaller than that for the e_g orbitals.

There are also some complexes with two metal–ligand distances *shorter* than the other four (Figure 5.19). For example, in the fluoride K_2CuF_4 there are two Cu–F distances of 195 pm and four Cu–F distances of 208 pm. Again the distorted state has an energy advantage over the octahedral situation with three electrons in e_g (Figure 5.23).

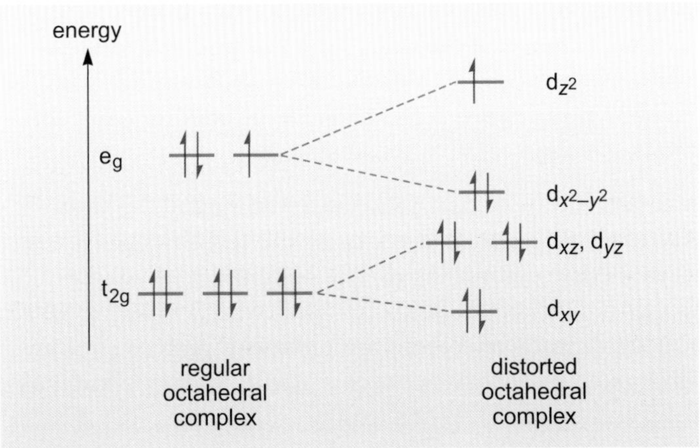

Figure 5.23 Splitting of the 3d levels for the distorted octahedral complex K_2CuF_4, in which two *trans* ligands on the *z*-axis are closer to the copper(II) ion than the four in the *xy*-plane.

The Jahn–Teller effect applies equally well to excited states as it does to ground-state d-electron configurations, as can be seen from the electronic spectra of some complexes. If you look back at Figure 5.11, you can see a small blip on the left side of the d–d band of the electronic spectrum of $[Ti(H_2O)_6]^{3+}$; this is called a *shoulder*. In a d^1 octahedral complex, the upper e_g level is degenerate and, in the ground (i.e. unexcited) state, is empty. In the excited state, e_g is occupied by a single electron and so splits into two levels to remove the degeneracy (Figure 5.24). Consequently, *two* closely spaced absorption bands, which overlap, arise because an electron in the d_{xz}, d_{yz} and d_{xy} levels can be promoted to either level.

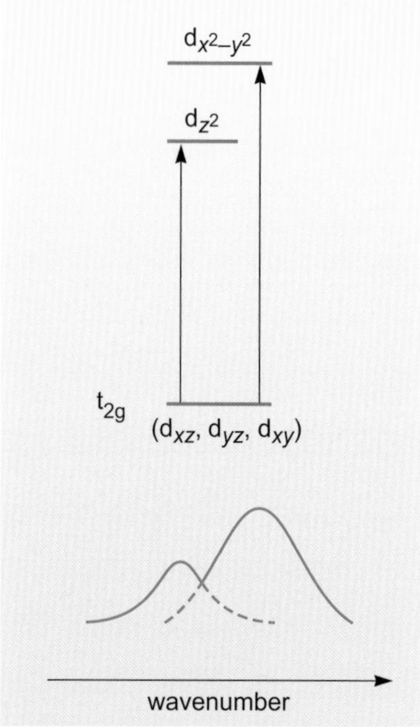

Figure 5.24 Energy-level diagram showing the electronic transitions in $[Ti(H_2O)_6]^{3+}$, taking into account splitting of the e_g level, and the corresponding part of the electronic spectrum. For simplicity, splitting of the t_{2g} level is not shown.

There is another, more striking example in Figure 5.25. This shows the absorption spectrum of $K_2Na[CoF_6]$, which contains the complex ion $[CoF_6]^{3-}$ (Co(III), d^6): two component bands, arising from the Jahn–Teller distortion of the $t_{2g}{}^3e_g{}^3$ excited state, are clearly defined.

5.6 Tetrahedral complexes

Tetrahedral complexes are found quite widely in transition-metal chemistry, and the bonding in these complexes can also be successfully treated using crystal-field theory. The easiest way to understand how the 3d levels split in a tetrahedral environment is to imagine a tetrahedral complex in a cube as in Figure 5.26, with the ligands occupying alternate corners; the *x*-, *y*- and *z*-axes run through the faces of the cube.

The crucial point to remember when applying crystal-field theory to tetrahedral complexes is that *no d orbital points directly at the ligands*, although some are closer than others. Look at the $3d_{xy}$ and $3d_{x^2-y^2}$ orbitals in Figure 5.27. Electrons in these orbitals are not equally repelled by the ligands of a tetrahedral complex. Although neither orbital points directly towards the ligands, an electron in $3d_{xy}$ is closer to the ligands and so is repelled more than an electron in $3d_{x^2-y^2}$.

Similarly, electrons in $3d_{xz}$ and $3d_{yz}$ orbitals are repelled more than one in $3d_{z^2}$. Hence, for a tetrahedral complex, the $3d_{xy}$, $3d_{xz}$ and $3d_{yz}$ orbitals are higher in energy than $3d_{x^2-y^2}$ and $3d_{z^2}$. The orbital energy-level diagram for a tetrahedral complex is shown in Figure 5.28. Like the octahedral case, the energy of *all* the d orbitals is raised relative to the free ion.

The lower doubly degenerate level ($3d_{x^2-y^2}$ and $3d_{z^2}$) in Figure 5.28c is labelled **e** and the higher triply degenerate level $\mathbf{t_2}$. (Note the absence of a g subscript here because tetrahedral complexes do not have a centre of symmetry.) The energy gap is labelled Δ_t (t for tetrahedral). As *none* of the d orbitals points directly at the ligands and there are only four ligands (not six as in octahedral), the difference in energy between e and t_2 is not as large as the difference between t_{2g} and e_g in octahedral complexes. In fact, if the metal ion, the ligands and the metal–ligand distance are the same in both octahedral and tetrahedral cases, it can be shown that $\Delta_t \approx \frac{4}{9}\Delta_o$. As a consequence, *virtually all tetrahedral complexes are high-spin* due to the smaller crystal-field splitting.

The CFSE in a tetrahedral complex, the *decrease in energy* on going from a spherical crystal field to a tetrahedral crystal field, may be calculated in exactly the same way as for octahedral complexes, only now each electron in

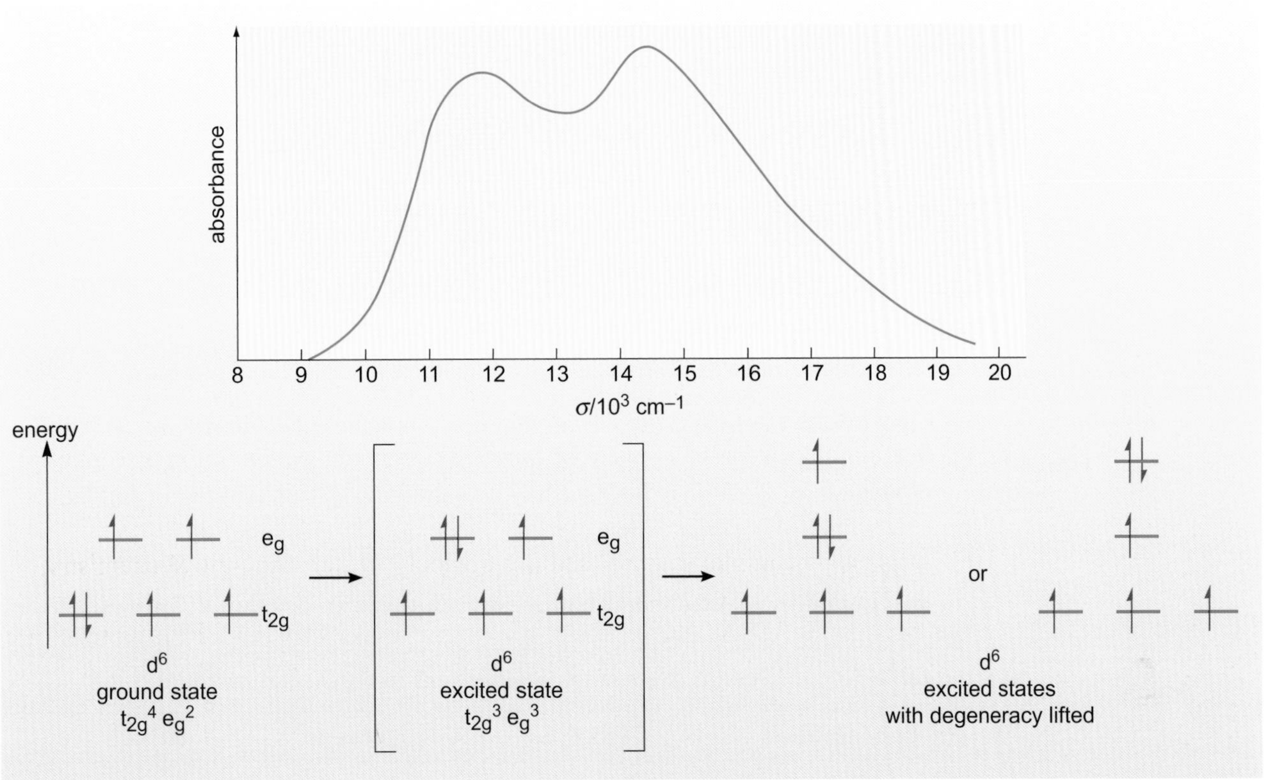

Figure 5.25 The electronic absorption spectrum of the $[CoF_6]^{3-}$ ion, showing the two peaks caused by the Jahn–Teller splitting of the excited state ($t_{2g}{}^3e_g{}^3$).

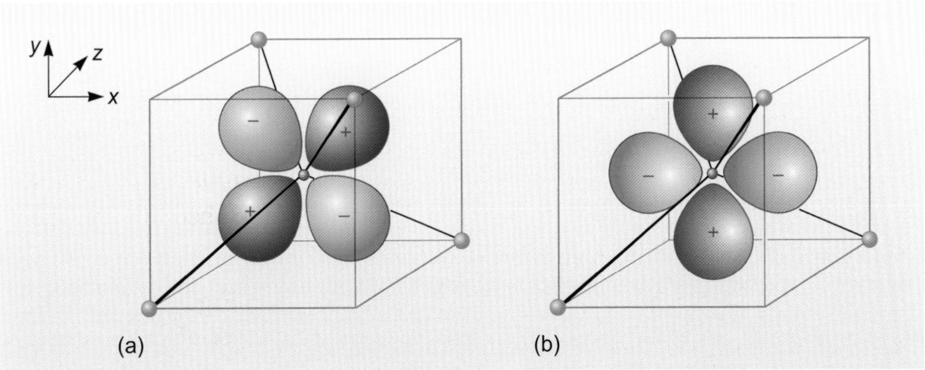

Figure 5.27 (a) A tetrahedral complex in a cube, showing the disposition of the ligands with respect to the metal $3d_{xy}$ orbital; (b) the same complex showing the position of the $3d_{x^2-y^2}$ orbital. Note the different perspective in this figure – you are looking along the *z*-axis.

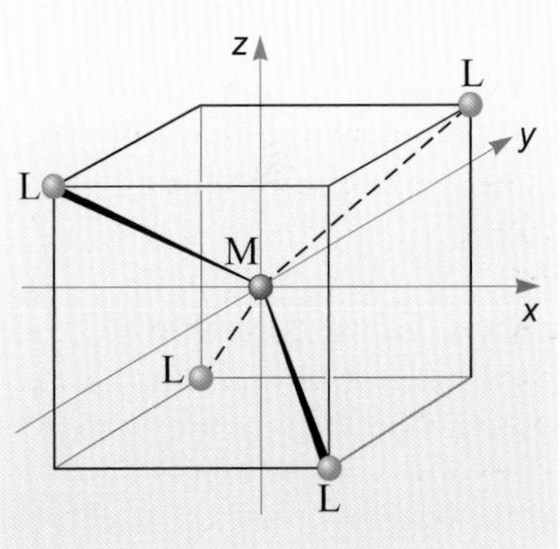

Figure 5.26 A tetrahedral complex in a cube, showing *x*-, *y*- and *z*-axes.

the e level contributes $\frac{3}{5}\Delta_t$ and each electron in the t_2 level contributes $-\frac{2}{5}\Delta_t$ to the CFSE.

The maxima in the tetrahedral CFSE occur at d^2 and d^7, which in part explains the occurrence of V^{3+} (d^2) tetrahedral complexes such as VX_4^- (where X = Cl, Br, I) and that cobalt(II) (d^7) forms more tetrahedral

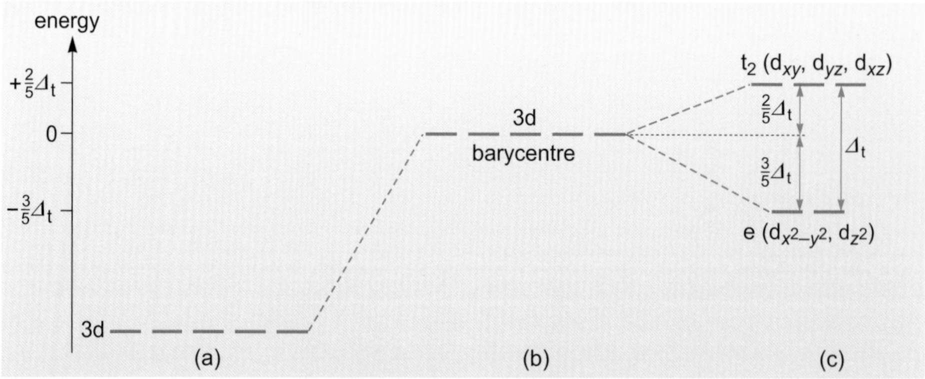

Figure 5.28 Orbital energy-level diagram (showing 3d orbitals only) for (a) a free ion, (b) an ion in a sphere of negative charge and (c) an ion in a tetrahedral complex.

complexes than any other transition-metal ion. The CFSEs for d^1–d^{10} configurations in tetrahedral complexes are given in Table 5.4.

Table 5.4 CFSEs for first transition series ions in tetrahedral complexes.

Configuration	CFSE*	Configuration	CFSE*
d^1	$\frac{3}{5}\Delta_t$	d^6	$\frac{3}{5}\Delta_t$
d^2	$\frac{6}{5}\Delta_t$	d^7	$\frac{6}{5}\Delta_t$
d^3	$\frac{4}{5}\Delta_t$	d^8	$\frac{4}{5}\Delta_t$
d^4	$\frac{2}{5}\Delta_t$	d^9	$\frac{2}{5}\Delta_t$
d^5	0	d^{10}	0

* Note that the CFSE defined here is the *decrease in energy* on going from a spherical crystal field to a tetrahedral crystal field. Thus the energy for an electron in an e level with a CFSE of $\frac{3}{5}\Delta_t$ is $-\frac{3}{5}\Delta_t$ relative to the barycentre (see Figure 5.28).

5.6.1 Colour in tetrahedral complexes

The colours of tetrahedral complexes are far more intense than their octahedral counterparts at the same concentration. This is because they do not possess a centre of symmetry, and so d↔d transitions are not forbidden by the rule g↮g.

Figure 5.29 shows the effect of adding concentrated hydrochloric acid to an aqueous solution of cobalt(II) chloride. The pale pink solution arises from the octahedral complex $[Co(H_2O)_6]^{2+}$, which is converted into the tetrahedral $[CoCl_4]^{2-}$ on addition of the acid:

$$\underset{\text{pink}}{[Co(H_2O)_6]^{2+}(aq)} + 4Cl^-(aq) \rightleftharpoons \underset{\text{deep blue}}{[CoCl_4]^{2-}(aq)} + 6H_2O(l) \quad (5.5)$$

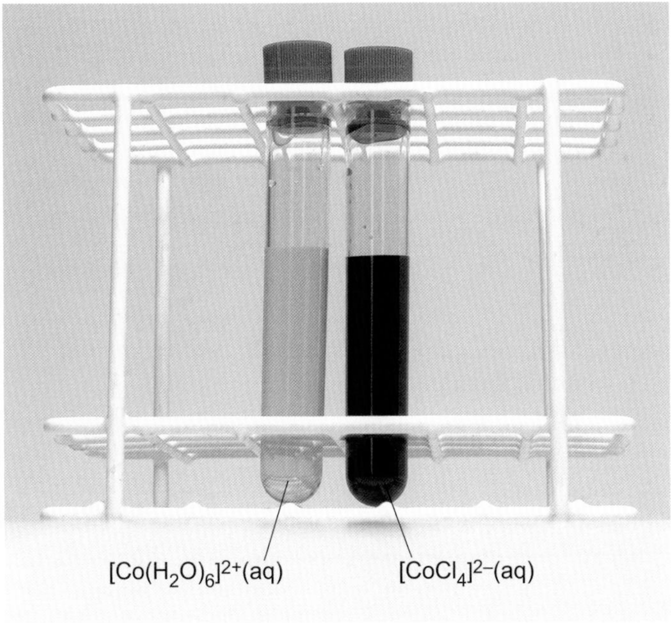

Figure 5.29 The pale pink solution (left-hand tube) was formed by dissolving cobalt(II) chloride in water. On adding concentrated hydrochloric acid, a deep blue solution (right-hand tube) containing the ion $[CoCl_4]^{2-}$ resulted.

A very common decorative colouring in glazes, and glass such as 'Bristol Blue', is the deep blue of Co^{2+} in a tetrahedral environment. The indicator used in the desiccant silica gel is also a tetrahedral Co^{2+} complex, which on absorbing moisture is converted from blue into a pale pink octahedral aquo complex.

Notice that tetrahedral $[CoCl_4]^{2-}$ is not only more intense in colour, but also absorbs at different wavelengths from the pale pink octahedral complex $[Co(H_2O)_6]^{2+}$. Similarly, in addition to being more strongly coloured, tetrahedral manganese(II) complexes are often green, whereas octahedral $[Mn(H_2O)_6]^{2+}$ is pale pink. The reason for this shift lies in the crystal-field splitting for tetrahedral complexes, Δ_t, which is smaller than Δ_o for octahedral complexes. Thus, the absorption bands in the spectra of the tetrahedral complex would be expected to be at lower energy (longer wavelength) – that is, further towards the red end of the visible spectrum – than those in the spectra of the octahedral complexes. A complex that absorbs in the red appears green. Conversely, octahedral complexes, which absorb higher energy, towards the green/blue end of the visible spectrum, appear red or, because the absorptions are very weak, pink.

5.6.2 Geometrical preferences of four-coordinate complexes

The most common stereochemistry for transition-metal complexes is octahedral, but towards the end of the transition-metal rows we find four-coordinate complexes. For the first row, these are most common in complexes of cobalt(II), nickel(II), copper(II), copper(I) and zinc(II). Table 5.5 shows the most common four-coordinate geometries for complexes of these metals.

Table 5.5 The most common four-coordinate geometries for first-row transition elements.

d^7 **cobalt(II)**	d^8 **nickel(II)**	d^9 **copper(II)**	d^{10} **copper(I)**	d^{10} **zinc(II)**
tetrahedral	square-planar and tetrahedral	square planar	tetrahedral	tetrahedral

The most common four-coordinate geometry is tetrahedral, where the four ligands arrange themselves to be as far apart as possible. For the d^7 and d^8 ions cobalt(II) and nickel(II), we saw earlier (Section 5.5.3) that square-planar complexes have an energy advantage over octahedral ones for strong-field ligands. For copper(II) (d^9), there is an energy advantage for square-planar rather than octahedral complexes with both strong- and weak-field ligands. If we compare the orbital energy-level diagram for tetrahedral and square-planar d^8 complexes (Figure 5.30), we can see that there is an energy advantage for square-planar over tetrahedral complexes. So we would expect square-planar geometry for copper(II) complexes and for low-spin (strong-field) cobalt(II) (d^7) and nickel(II) (d^8) complexes on crystal-field grounds.

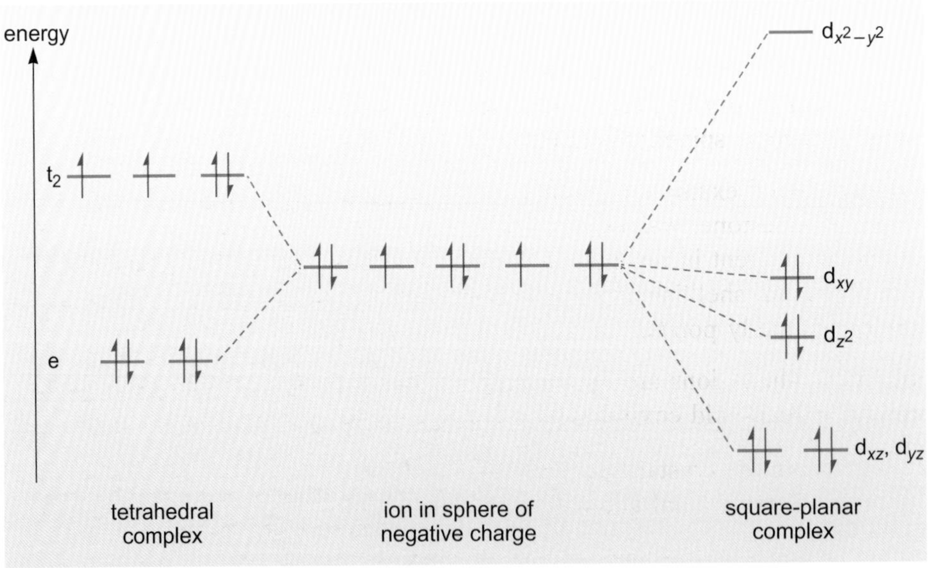

Figure 5.30 Energy-level diagrams for d^8 tetrahedral and square-planar complexes.

■ Co in the unusual oxidation state of +1 is found in cobalamin (a molecule related to vitamin B_{12}). What is the d electron configuration of Co(I) and what geometry would you expect cobalamin to adopt around the Co^+ ion?

□ Co^+ is a d^8 ion. It would be expected to favour a square-planar geometry. The geometry around Co in cobalamin is indeed found to be roughly square-planar.

Other factors also affect the geometry, particularly steric constraints imposed by multidentate ligands; for example, very large ligands may force a complex to be tetrahedral because the larger ligand–metal–ligand angle affords more space. On the other hand, the structures of some bidentate ligands, such as 1,2-diaminoethane (en) may be such that the distance between the two coordinating atoms fits a square plane better than a tetrahedron. Biological systems are particularly adept at tailoring ligands to force the ion to adopt a particular geometry.

Note that no crystal-field splitting labels (Δ) are shown on the d-orbital energy-level diagram for a square-planar complex. This is because there are several possible energy-level differences. However, the difference in energy between the d_{xy} and $d_{x^2-y^2}$ is the same as Δ_o if the complex is made of the same metal and ligands in both cases.

5.6.3 Site occupation in spinels

Crystal-field theory can provide information about the preferred location of metal ions in minerals. One example is the class of oxides known as spinels.

Spinels, such as the oxide $MgAl_2O_4$, have the general formula AB_2O_4 where A is a divalent cation and B a trivalent cation. The structure is based on a close-packed array of oxide ions with A ions occupying tetrahedral holes and B ions octahedral holes. In inverse spinels, the A ions occupy octahedral holes and the B ions are equally divided between octahedral and tetrahedral holes. Crystal-field theory can tell us whether a particular compound will adopt the spinel or inverse spinel structure.

A widely found example of such a crystal structure is magnetite, Fe_3O_4, which, as lodestone, was used in earlier times as a compass and is used today as a contrast agent in magnetic resonance imaging (MRI). It is found in the teeth of certain shellfish and is used by magnetotactic bacteria to locate the bottom of muddy ponds.

For Fe_3O_4, the A ions are Fe^{2+} and the B ions Fe^{3+}. The oxide ions will provide a weak-field environment.

■ What are the crystal-field stabilisation energies (CFSEs) of Fe^{2+} and Fe^{3+} in weak tetrahedral and octahedral fields?

□ In octahedral sites the CFSEs are $Fe^{2+}(d^6)$: $\frac{2}{5}\Delta_o$, $Fe^{3+}(d^5)$: 0. In tetrahedral sites the CFSEs are $Fe^{2+}(d^6)$: $\frac{3}{5}\Delta_t$, $Fe^{3+}(d^5)$: 0.

Remembering that $\Delta_t \approx \frac{4}{9}\Delta_o$, this means that the structure will be stabilised if Fe^{2+} ions occupy octahedral sites. There are only enough Fe^{2+} ions to fill half the available octahedral holes and so the Fe^{3+} ions are split between octahedral and tetrahedral sites. The structure of Fe_3O_4 (Figure 5.31) with A ions in octahedral positions and B ions spread between octahedral and tetrahedral positions is an inverse spinel structure.

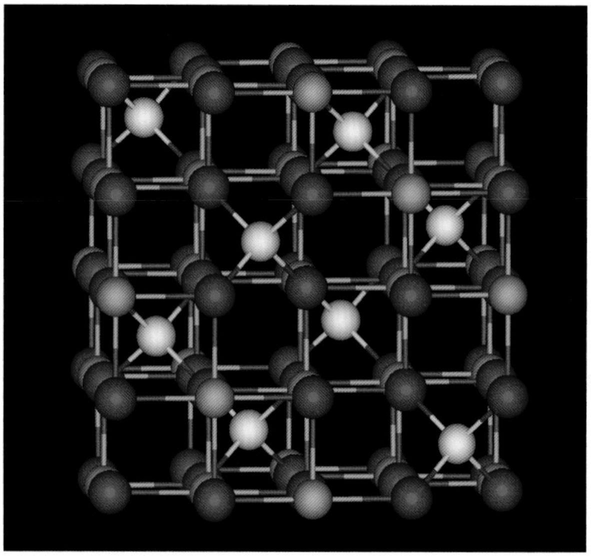

Figure 5.31 Structure of magnetite (red spheres O, orange spheres Fe on octahedral sites, lemon spheres Fe on tetrahedral sites.)

5.7 The magnetic properties of transition-metal complexes

One way of determining whether a complex is high-spin or low-spin is to establish the number of unpaired electrons. In this section, we shall see how measurements of the magnetism of transition-metal complexes can do this.

A magnetic field is generated by a moving charge. **Diamagnetism** occurs in all substances and is a weak effect caused by the circulation of electrons in atoms and molecules. A magnetic field induces electrons to circulate producing a magnetic moment in the opposite direction to the field. Diamagnetic substances are thus *repelled weakly* by a magnetic field, but the effect is so small, it is usually swamped by other types of magnetism.

Paramagnetic substances are *attracted* to a magnetic field. **Paramagnetism** is due to isolated unpaired spins of electrons on atoms or molecules. The set of unpaired electrons in one complex does not interact with the set in a neighbouring complex because the complexes are too far apart; the different sets of unpaired electrons are thus oriented randomly with respect to one another and are said to be *magnetically dilute*.

If unpaired spins are coupled to each other so that they line up and reinforce one another, then the material is said to be **ferromagnetic**. **Ferromagnetism** occurs in iron (from which it takes its name) and cobalt, and it is this much stronger property that is generally referred to as 'magnetism' in everyday life. In **antiferromagnetic** substances, the spins line up and interact with one another in such a way that they cancel each other out.

In oxides and other solids, metal ions can occupy sites with different geometries around them. In this case, we have to consider coupling between spins on similar sites and coupling between spins on different sites. This can lead to partial cancellation. If the net result is that more of the spin coupling is reinforcing, then the substance is **ferrimagnetic**. The classic example of this is magnetite, Fe_3O_4. The ions on the tetrahedral sites have spins opposed to those on the octahedral sites. As half the Fe^{3+} ions occupy tetrahedral sites and half occupy octahedral sites (see Section 5.6.3), there is no net magnetisation from the Fe^{3+} ions. This leaves the Fe^{2+} ions whose spins are all parallel as they all occupy octahedral sites. Thus Fe_3O_4 is ferrimagnetic.

For many transition-metal complexes, however, the metal ions are sufficiently far apart that we can ignore coupling between them and they can be described as paramagnetic or diamagnetic. It is measurements on such complexes that we are concerned with here.

5.7.1 Paramagnetism

Magnetic fields are produced by circulating charge. The *magnetic field strength*, H, is proportional to the density of the lines of magnetic force, the *magnetic flux density*, B. In a vacuum, B and H are related by a constant called the permeability of free space, μ_0:

$$B = \mu_0 H \tag{5.6}$$

B is measured in tesla, T (tesla is the SI unit of magnetic field intensity; 1 T is a very strong magnetic field), and H in amperes per metre, A m^{-1} (an ampere is the SI unit of electric current); μ_0 has the value $4\pi \times 10^{-7}$ T m A^{-1}.

If a substance is placed in the magnetic field rather than a vacuum, it contributes its own field. The total field strength is then increased or decreased by the field strength of the sample, which is called the **magnetisation**, $\boldsymbol{M}$. This can be expressed by:

$$B = \mu_0(H + M) \tag{5.7}$$

The magnetisation, M, is proportional to the magnetic field strength H:

$$M = \chi H \tag{5.8}$$

The quantity χ (Greek letter chi, pronounced 'kye') is known as the **magnetic susceptibility**. Substituting and rearranging allows us to rewrite Equation 5.7:

$$B = \mu_0 H(1 + \chi) \tag{5.9}$$

χ is negative for diamagnetic materials and positive for paramagnetic materials. Diamagnetic substances are thus repelled weakly by an inhomogeneous magnetic field, whereas paramagnetic substances are attracted.

Because all substances display diamagnetism, the magnetic susceptibility of paramagnetic substances contains a contribution from both diamagnetism and paramagnetism, although the paramagnetism is dominant. Both diamagnetic and paramagnetic susceptibilities are independent of the magnetic field strength, H.

Experimentally, it was shown by Pierre Curie that the paramagnetic susceptibility, χ^{para}, is proportional to the inverse of the temperature, the **Curie law** (Figure 5.32):

$$\chi^{para} = \frac{C}{T} \tag{5.10}$$

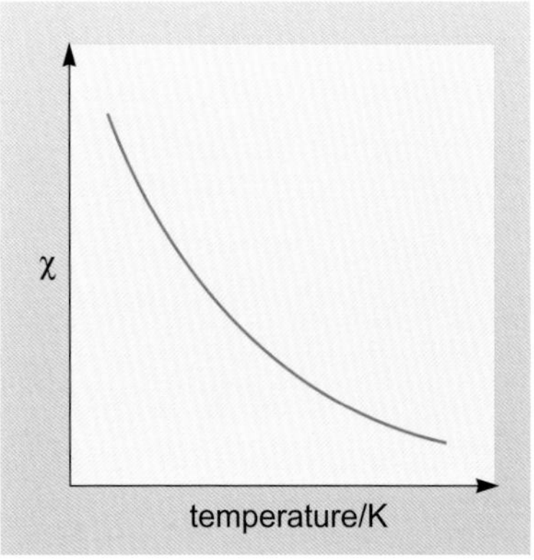

Figure 5.32 The temperature dependence of paramagnetic susceptibility.

where C is known as the **Curie constant**, and is characteristic of the complex, and T is temperature.

To understand why a paramagnetic sample behaves in this way, it is useful to think of the sample as a collection of transition-metal complex ions, each of which behaves as a tiny magnet. An applied magnetic field tends to make them line up with the field, but when the temperature is above 0 K, they also have thermal energy, and this tends to make them move around. This movement alters their orientation so that they are no longer lined up with the field but are more randomly distributed in direction. As the temperature rises, the thermal energy increases, the magnets move more and become more randomly orientated. Thus, as shown by Equation 5.10, the magnetic susceptibility varies, but the value of the magnetic field produced by one of these little magnets remains constant with temperature. It is these magnetic fields, known as **magnetic moments, μ**, which give us chemically useful information, and their value can be obtained from the Curie constant, C, for that particular substance. Magnetic moments are usually quoted in the unit **Bohr magneton, μ_B**, given by:

$$1\,\mu_B = \frac{eh}{4\pi mc} = 9.274\times10^{-24}\ \mathrm{J\,T^{-1}} \qquad (5.11)$$

where e is the charge on an electron, h is Planck's constant, m is the mass of an electron and c is the speed of light. We now need to relate the magnetic moment to the number of unpaired electrons in a complex.

5.7.2 Spin magnetic moments

An electron is a charged particle and circulating charges produce a magnetic field. All electrons have spin and electron spin thus gives rise to a magnetic field. The magnetic moment, μ_s, of a *single* free electron is given by the equation:

$$\mu_s = g\sqrt{s(s+1)}\ \mu_B \qquad (5.12)$$

where s is the value of the spin quantum number ($\frac{1}{2}$); $\sqrt{s(s+1)}$ is the magnitude of the spin angular momentum of the electron; and g is the **gyromagnetic ratio** or ***g* factor** which for the free electron has the value 2.00023. (Because magnetic moments cannot be measured this accurately, it is sufficient to use the value 2.)

Substituting $s = \frac{1}{2}$ in Equation 5.12, we calculate the spin magnetic moment of a free electron as 1.73 μ_B. Thus we would expect that any transition-metal complex with a d^1 configuration would have a magnetic moment of 1.73 μ_B. In practice, they are usually close to this value, but it can be changed a little by orbital effects, as we shall see in Sections 5.7.3–5.7.5.

For a complex, the **spin magnetic moment, μ_S**, is given by:

$$\mu_S = g\sqrt{S(S+1)}\ \mu_B \qquad (5.13)$$

where S is the total spin quantum number of the atom and is simply the sum of the spin quantum numbers, $s = \frac{1}{2}$, for each of the electrons in the complex.

■ What is the maximum number of unpaired spins found in transition-metal complexes?

□ The answer is five. There are five d orbitals, and the maximum number of unpaired spins is when there is one electron in each d orbital. Try calculating S and μ_S for one to five unpaired electrons. The results are given in Table 5.6.

Table 5.6 Values of S and μ_S for one to five unpaired electrons.

Number of unpaired electrons	S	μ_S
1	$\frac{1}{2}$	$2\sqrt{\frac{1}{2}\times\frac{3}{2}} = 1.73\,\mu_B$
2	1	$2\sqrt{1\times 2} = 2.83\,\mu_B$
3	$\frac{3}{2}$	$2\sqrt{\frac{3}{2}\times\frac{5}{2}} = 3.87\,\mu_B$
4	2	$2\sqrt{2\times 3} = 4.90\,\mu_B$
5	$\frac{5}{2}$	$2\sqrt{\frac{5}{2}\times\frac{7}{2}} = 5.92\,\mu_B$

Now turning the argument on its head, assuming that the magnetic moment is due to electron spin only, the number of unpaired electrons in a complex can be determined from its measured value. In terms of the number of unpaired electrons, n, Equation 5.13 can be rewritten as:

$$\mu_S = \sqrt{n(n+2)}\,\mu_B \qquad (5.14)$$

This is known as the **spin-only formula**.

We noted in Section 5.5.3 that Fe_3O_4 was used as a contrast agent in MRI. Some metal complexes are also used for this purpose. A high value of the magnetic moment is one criterion in determining suitability for this application.

■ One complex used as a contrast agent contains Mn(II). Suggest why Mn(II) is a good choice.

□ Mn(II) complexes have the electronic configuration d^5. In high-spin complexes all the d electrons are unpaired and the magnetic moment is close to 5.92 μ_B.

Magnetic measurements can thus be used to identify weak-field and strong-field complexes. Taking the example of an octahedral cobalt(II) complex ion with a magnetic moment of 1.92 μ_B. How do we work out the electronic configuration of the metal ion? Cobalt(II) complexes have seven 3d electrons. In a weak-field octahedral environment, these would give a configuration $t_{2g}^{\ 5}e_g^{\ 2}$, with three unpaired electrons and we would expect a magnetic moment of 3.87 μ_B. In a strong field, the configuration would be $t_{2g}^{\ 6}e_g^{\ 1}$, with only one unpaired electron. Because 1.92 μ_B is much closer to the spin magnetic moment for one unpaired electron (1.73 μ_B), we can conclude that this is a strong-field complex.

■ What would you expect to be the magnetic moment of the complex $[CoF_6]^{3-}$?

□ First, decide how many unpaired electrons you would expect. Co^{3+} is a d^6 ion, and F^- is a weak-field ligand and thus gives a high-spin octahedral complex with four unpaired electrons. We would expect a magnetic moment of 4.9 μ_B.

■ A class of enzymes known as iron-containing nitrile hydratase contain Fe(III) in a near-octahedral coordination. The ligating atoms around the Fe include three S atoms (from cysteine). Would you expect the Fe(III) ion in the enzyme to be in a high-spin or low-spin state?

□ S-ligating ligands generally lead to high-spin states because of the position of such ligands in the spectrochemical series.

■ The measured value of S for the complex is $\frac{1}{2}$. What is the actual electronic configuration of Fe in this complex?

□ Fe^{3+} is a d^5 ion and in the high-spin state has five unpaired electrons and thus $S = \frac{5}{2}$. A value of $S = \frac{1}{2}$ suggests a low-spin state (t_{2g}^5).

Comment: Studies on model compounds indicate that in this case the geometry of the ligands prevents ligand–metal orbital interactions that would lead to a weak field. These interactions will be covered in Chapter 6 when you study the molecular orbital theory of complexes.

5.7.3 Deviation from the spin-only formula

Table 5.7 shows the observed magnetic moments for octahedral (weak-field) and tetrahedral complexes for common first-row transition-metal ions.

As you can see, many of the observed values are close to μ_S, but some ions, particularly d^6, d^7 and tetrahedral d^8 ions, have larger magnetic moments than expected. We need to find an explanation for this. So far, we have only considered contributions to the magnetic moment from the electron spin.

■ If we think of an analogy for electron spin of the Earth spinning on its axis, what other type of roughly circular motion does the Earth undergo?

□ The Earth goes round the Sun; this is *orbital motion*.

Table 5.7 Observed magnetic moments for octahedral (weak-field) and tetrahedral complexes for first-row transition-metal ions.

Number of d electrons		Predicted spin-only magnetic moment	Octahedral complex		Tetrahedral complex	
Total	Unpaired	μ_S/μ_B	Configuration	μ_{obs}/μ_B	Configuration	μ_{obs}/μ_B
1	1	1.73	$t_{2g}{}^1$	1.7–1.8	e^1	≈1.7
2	2	2.83	$t_{2g}{}^2$	2.8–2.9	e^2	2.6–3.0
3	3	3.87	$t_{2g}{}^3$	3.7–3.9	$e^2t_2{}^1$	—
4	4	4.90	$t_{2g}{}^3e_g{}^1$	4.8–5.0	$e^2t_2{}^2$	—
5	5	5.92	$t_{2g}{}^3e_g{}^2$	5.8–6.0	$e^2t_2{}^3$	5.8–6.0
6	4	4.90	$t_{2g}{}^4e_g{}^2$	5.1–5.7	$e^3t_2{}^3$	5.0–5.2
7	3	3.87	$t_{2g}{}^5e_g{}^2$	4.3–5.2	$e^4t_2{}^3$	4.4–4.8
8	2	2.83	$t_{2g}{}^6e_g{}^2$	2.8–3.4	$e^4t_2{}^4$	3.7–4.0
9	1	1.73	$t_{2g}{}^6e_g{}^3$	1.7–2.2	$e^4t_2{}^5$	1.75–2.2
10	0	0	$t_{2g}{}^6e_g{}^4$	0	$e^4t_2{}^6$	0

5.7.4 The orbital magnetic moment

We can think of an electron in an atom or an ion going round the nucleus, although the analogy is not perfect as electrons do not travel around a nucleus in a fixed orbit like the Earth round the Sun. The electron thus has orbital motion too, and this gives rise to an **orbital magnetic moment**. We have to combine the spin *and* orbital contributions to obtain the total magnetic moment:

$$\mu_{S+L} = \sqrt{g^2S(S+1) + L(L+1)}\ \mu_B \qquad (5.15)$$

where S is the total spin quantum number of the atom, and L is the quantum number for an atom defining the total orbital contribution.

Box 5.3 Quantum numbers of atoms

n = 1, 2, 3 … is the principal quantum number (the shell number); the orbital momentum quantum number $l = n-1, n-2 \ldots 0$ defines the specific sub-shells, s, p, d, etc.; the magnetic quantum number $m_l = 0, \pm1, \pm2, \ldots \pm l$ distinguishes the individual orbitals within a sub-shell; and m_s is the magnetic spin quantum number, $m_s = \pm\frac{1}{2}$ distinguishing spin up from spin down. Every electron in the atom must have its own unique set of these four quantum numbers. The spin quantum number, s, is always $\frac{1}{2}$ for a single electron.

We found the total spin quantum number S by adding $\frac{1}{2}$ for every unpaired electron, but for orbital angular momentum, things are not so simple. An electron in a d orbital has an orbital quantum number, $l = 2$. For a d^2 atom, however, L is the combined quantum number for both electrons. No two electrons on an atom or ion can have the same set of four quantum numbers (n, l, m_l and m_s). If we are considering an ion from the first-row transition elements, its 3d electrons have the same principal quantum number, $n = 3$ and orbital quantum number, $l = 2$. All the unpaired parallel electrons must have the same spin quantum number, say, $m_s = \frac{1}{2}$, so the variation must come in the magnetic quantum number m_l for which there are five possible values, namely +2, +1, 0, −1 or −2. If there are two d electrons, they must have different values of m_l. The value of L is found by taking the maximum value of m_l for the two electrons. The largest value for a single electron is +2, but if one electron has +2, the other electron must have one of the other values. The maximum value left is +1. For d^2, L is thus $(2 + 1) = 3$. For d^3, the third electron has to have $m_l = 0$ for the maximum value, and so $L = (2 + 1 + 0) = 3$ for d^3 as well.

■ What are the values of L for d^4 and d^5 configurations with all spins unpaired and parallel?

□ For d^4, $L = 2 + 1 + 0 + (-1) = 2$ and,

for d^5, $L = 2 + 1 + (0) + (-1) + (-2) = 0$

■ Now calculate the total magnetic moment, m_{S+L}, for high-spin ions of configuration d^1 to d^5 using Equation 5.15.

□ The values are given in Table 5.8 and compared with the spin-only values and the experimentally measured values for transition-metal complexes.

Table 5.8 Theoretical total, spin-only and observed magnetic moments.

Number of unpaired electrons	S	L	μ_{S+L}/μ_B	μ_S/μ_B	μ_{obs}/μ_B
1	$\frac{1}{2}$	2	3	1.73	1.7–1.8
2	1	3	4.47	2.83	2.6–3.0
3	$\frac{3}{2}$	3	5.20	3.87	3.7–3.9
4	2	2	5.48	4.90	4.8–5.0
5	$\frac{5}{2}$	0	5.92	5.92	5.8–6.0

Comparing the observed values in the final column with the theoretical values, we see that adding in the full orbital contribution like this is obviously an overestimate. It would appear, then, that there is some orbital contribution but not as much as implied by Equation 5.15, which applies to the free ions.

5.7.5 Magnetic moments in complexes

In a *free* ion, the electron can circulate around a field along the z-axis by going from d_{xz} to d_{yz} (which are combinations of orbitals with $m_l = +1$ and $m_l = -1$) or by going from $d_{x^2-y^2}$ to d_{xy} (which are combinations of orbitals with $m_l = +2$ and $m_l = -2$). An electron in d_{z^2} ($m_l = 0$) has no other orbital to move into with the same value of m_l, and does not contribute to the orbital magnetic moment. However, here we are concerned with ions in complexes where the d orbital energies are no longer degenerate but are split by the crystal field. If the ion is in an octahedral or tetrahedral complex, then $d_{x^2-y^2}$ and d_{xy} no longer have the same energy. The only orbital contribution in these complexes thus comes from d_{xz} and d_{yz}, which are degenerate: in octahedral complexes, d_{xz} and d_{yz} belong to t_{2g}, and in tetrahedral complexes to t_2. If both orbitals are occupied by electrons of the same spin, then one has $m_l = +1$ and one $m_l = -1$, and the magnetic moments cancel. An orbital contribution to the magnetic moment is therefore only expected if there is a vacancy in d_{xz} or d_{yz}. Thus, we only expect orbital contributions for complexes with one, two, four and five electrons in t_{2g} or t_2 (if the occupation is three or six electrons, then the orbitals are either all singly occupied or completely occupied).

■ Which of the configurations d^1 to d^9 for weak-field octahedral and tetrahedral complexes are expected to show an orbital contribution to the magnetic moment?

□ Octahedral: d^1 ($t_{2g}{}^1$), d^2 ($t_{2g}{}^2$), d^6 ($t_{2g}{}^4e_g{}^2$), d^7 ($t_{2g}{}^5e_g{}^2$);

tetrahedral: d^3 ($e^2t_2{}^1$), d^4 ($e^2t_2{}^2$), d^8 ($e^4t_2{}^4$), d^9 ($e^4t_2{}^5$)

These expectations are generally borne out, except that there are no data for d^3 and d^4 tetrahedral complexes.

In general, magnetic measurements are not useful in distinguishing between weak-field tetrahedral and octahedral environments, as both have the same number of unpaired electrons. There is some correlation between geometry and magnetic moment, however, for complexes of cobalt(II) and nickel(II), due to the differing orbital magnetic moment contributions.

Cobalt(II) complexes are d^7, so the weak-field complexes have three unpaired electrons and we expect a spin-only value of $\mu_S = 3.87\ \mu_B$. In practice, the octahedral complexes have an orbital contribution and higher magnetic moments (4.7–5.2 μ_B) than the tetrahedral complexes (4.4–4.8 μ_B).

For nickel(II), d^8, there are only two unpaired electrons, we expect a spin-only value of $\mu_S = 2.83\ \mu_B$ and it is the tetrahedral complexes that have higher orbital contributions, with magnetic moments in the range 3.7–4.0 μ_B, compared with 2.83–3.4 μ_B for octahedral complexes.

The observed values for tetrahedral nickel(II) and octahedral cobalt(II) complexes show that care must be taken in interpreting the results if large orbital contributions are expected, as the values for these compounds are close to the spin-only values for complexes with one more unpaired electron.

So far, we have only considered tetrahedral and octahedral complexes. In substituted and slightly distorted complexes, although we lose the degeneracy of some of the 3d levels, we generally have the same number of unpaired electrons. For predicting the spin-only magnetic moment, therefore, we can regard such complexes as tetrahedral or octahedral. In some cases, the degeneracy of d_{xz} and d_{yz} is lost, and the magnetic moment is then closer to the spin-only value than that of true tetrahedral or octahedral complexes.

Don't forget that there are questions on the companion website which you can use to test your understanding of the material covered in this chapter.

6 Molecular orbital theory of transition-metal complexes

As you have seen in Chapter 5, simple crystal-field theory can be used to explain some of the properties of transition-metal complexes, but there are occasions when we need a more sophisticated theory. For example, the position of ligands in the spectrochemical series was purely empirical; we had no satisfactory explanation for the order. Indeed, we noted in Section 5.4 that a simple electrostatic model led to a prediction at odds with the empirical order of ligands in the series. Again, we noted that many of the deepest colours found for transition-metal complexes are not due to d↔d transitions, but to transitions between metal-based and ligand-based orbitals.

In crystal-field theory, we regarded the ligands as negative point charges and considered the effect of these on the metal 3d atomic orbitals. In **molecular orbital theory**, we form *molecular* orbitals for the entire complex by combining the metal 3d orbitals with orbitals on the ligands. This approach not only deals with some of the problems noted for crystal-field theory, but also is the basis for accurate calculations of properties of transition-metal complexes.

Advances in computer hardware in the last decades of the 20th century mean that *ab initio* methods (that is, molecular orbital methods that solve the Schrödinger equation without using information from experiments) can be applied to transition-metal complexes. Such calculations are used not only to complement experimental studies on simple transition-metal complexes but also to provide insights into, for example, the active sites of metallo-enzymes from predicted properties of model complexes.

We now look at what type of molecular orbitals we find for transition-metal complexes. We start by reminding you how molecular orbitals are formed for small molecules containing only atoms of the main-Group elements.

For these molecules, the orbitals are built up using atomic orbitals on the atoms in the molecule. To decide which atomic orbitals to combine, we use the following guidelines:

- atomic orbitals that combine must be of similar energy
- only atomic orbitals of the same symmetry can combine
- there must be significant overlap of combining orbitals
- *n* atomic orbitals combine to make *n* molecular orbitals.

For transition-metal complexes, we need to decide which orbitals on the ligands combine with the d orbitals on the metal. For simple complexes, such as $[MCl_6]^{n-}$, we consider the valence atomic orbitals on the ligand atom that are closest in energy to the metal d orbitals. However, for most complexes the ligands are not single atoms but molecules. The general strategy here is first to form molecular orbitals for the individual ligand molecules, and then consider which of these molecular orbitals combine with the metal d orbitals. The ligand molecular orbitals that are of the correct energy to combine with metal d orbitals are usually the highest-occupied and lowest-unoccupied orbitals. For example, Figure 6.1 shows a partial molecular orbital energy-level diagram for CO. The valence electrons have been fed into the lowest energy orbitals in pairs. 5σ is known as the **HOMO**, the highest occupied molecular orbital, and 2π* as the **LUMO**, the lowest unoccupied molecular orbital.

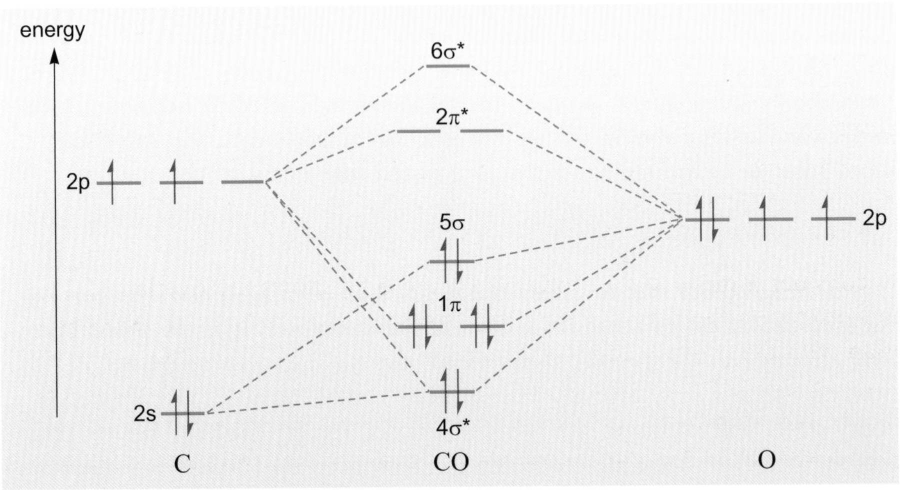

Figure 6.1 Partial orbital molecular energy-level diagram for CO.

To form orbitals for a transition-metal complex with CO ligands (called a carbonyl complex), we would consider the occupied 5σ and 1π orbitals, and the unoccupied 2π* and 6σ* orbitals formed from the C2p and O2p atomic orbitals. Ligand orbitals can be divided into two sets – σ-bonding orbitals, which overlap with the metal orbital to increase electron density along the metal–ligand bond, and π-bonding orbitals, where the electron density is above and below the bond. We start by considering just the σ-bonding orbitals.

6.1 σ-bonding

Examples of ligand orbitals which form σ orbitals with metal d orbitals are s and p orbitals on halide ions, σ orbitals on diatomic ligands such as OH^- and CO, and bonding molecular orbitals formed from H 1s and N or O 2p atomic orbitals in ligands such as H_2O and NH_3. We represent a general **σ-bonding orbital** by a tear-shape as in Figure 6.2. Filled orbitals of this type are usually of lower energy than the metal d orbitals, and so we will assume this in the subsequent discussion.

Figure 6.2 A generalised ligand σ-bonding orbital.

We shall start with an octahedral complex and see how we can build up a molecular orbital diagram for such a complex. Think of a metal ion surrounded by six ligands, each contributing one filled orbital. Thus, there are the five metal d orbitals and six ligand orbitals from which to construct molecular orbitals for the complex. Now arrange the six ligands to lie on the *x*-, *y*- and *z*-axes and look at how the six ligand orbitals overlap with the metal d orbitals: Figure 6.3 illustrates this for a σ-bonding ligand orbital on the *z*-axis overlapping with the d_{z^2} orbital on the metal.

■ Would such a σ-bonding orbital combine with the d_{z^2} orbital?

□ Yes. The orbitals have lobes of the same sign overlapping, and so can form a **bonding orbital**.

The other ligand orbital on the *z*-axis also overlaps in the same way. The ligand orbitals along the *x*- and *y*-axes overlap with the torus, but now have the opposite sign. So the metal d_{z^2} orbital can combine with orbitals on all six ligands. In other words, there is a combination of the six σ-bonding ligand orbitals that has the same symmetry as the metal d_{z^2} orbital. This combination of six ligand orbitals with the metal d_{z^2} orbital must produce two orbitals for the complex – one bonding and one antibonding. Figure 6.4 shows this combination of ligand orbitals, and its interaction with the metal d_{z^2} orbital to form the **bonding** and **antibonding orbitals** of the complex.

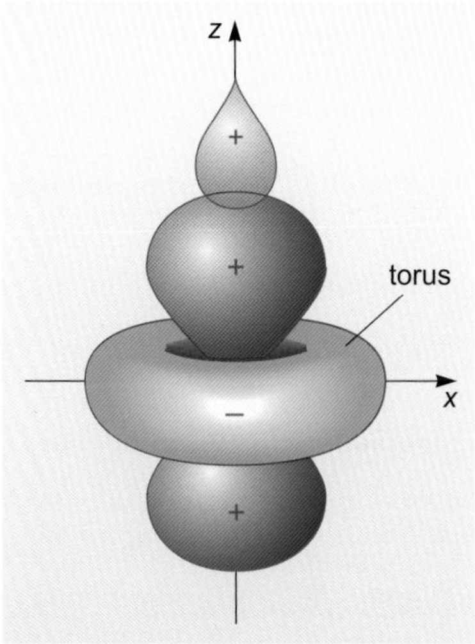

Figure 6.3 Overlap of a ligand σ-bonding orbital with the d_{z^2} orbital on the metal atom.

Now consider a metal $d_{x^2-y^2}$ orbital. Figure 6.5a shows the overlap of a ligand orbital on the *z*-axis. Here there is no net overlap because the bonding overlap with the positive lobes is cancelled out by the antibonding overlap with the negative lobes. Ligand orbitals of the appropriate sign, however, do have a net overlap with the metal $d_{x^2-y^2}$ orbital on the *x*- and *y*-axes (Figure 6.6). The $d_{x^2-y^2}$ orbital, therefore, can combine with a combination of four ligand orbitals of the same symmetry as itself. Again, one bonding and one antibonding orbital are formed from the $d_{x^2-y^2}$ orbital and this combination.

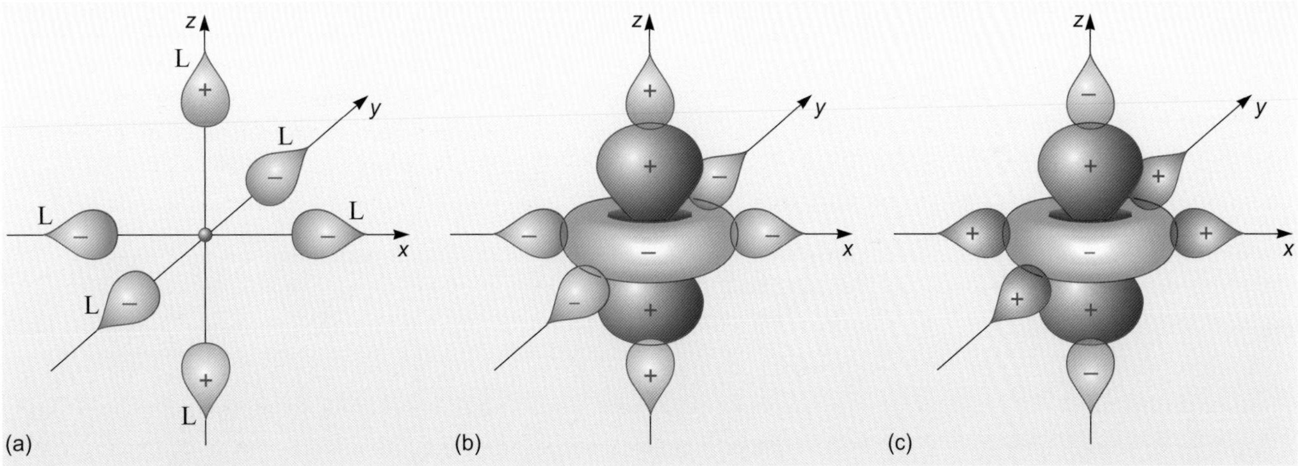

Figure 6.4 (a) Combination of six σ-bonding ligand orbitals around a metal atom; (b) the bonding orbital for the complex formed from the ligand orbital combination and the metal d_{z^2} orbital; and (c) the corresponding antibonding orbital for the complex.

Figure 6.6 shows the combination of four ligand orbitals overlapping with the metal $d_{x^2-y^2}$ orbital to form (a) a bonding orbital and (b) an antibonding orbital.

Thus, d_{z^2} and $d_{x^2-y^2}$ orbitals on the metal overlap with σ-bonding orbitals on the ligands to form one bonding and one antibonding orbital each for the metal complex.

Calculations of the energies of the bonding orbitals formed from the metal d_{z^2} and $d_{x^2-y^2}$ orbitals show that the complex bonding orbitals are also degenerate. These are labelled e_g, the label for d_{z^2} and $d_{x^2-y^2}$ orbitals in an octahedral environment. The antibonding complex orbitals must also be degenerate; they are labelled e_g*.

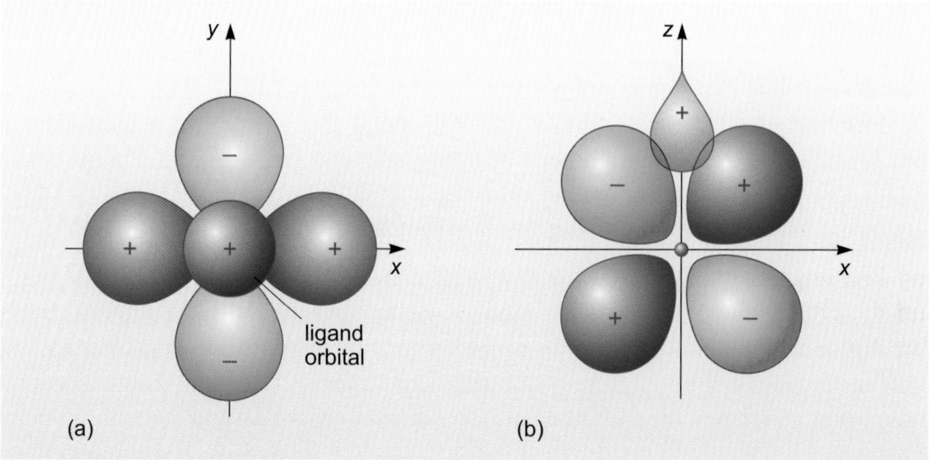

Figure 6.5 (a) A combination of a σ-bonding ligand orbital on the z-axis and a metal $d_{x^2-y^2}$ orbital results in no net overlap. (b) A σ-bonding ligand orbital on the z-axis and a metal d_{xz} orbital.

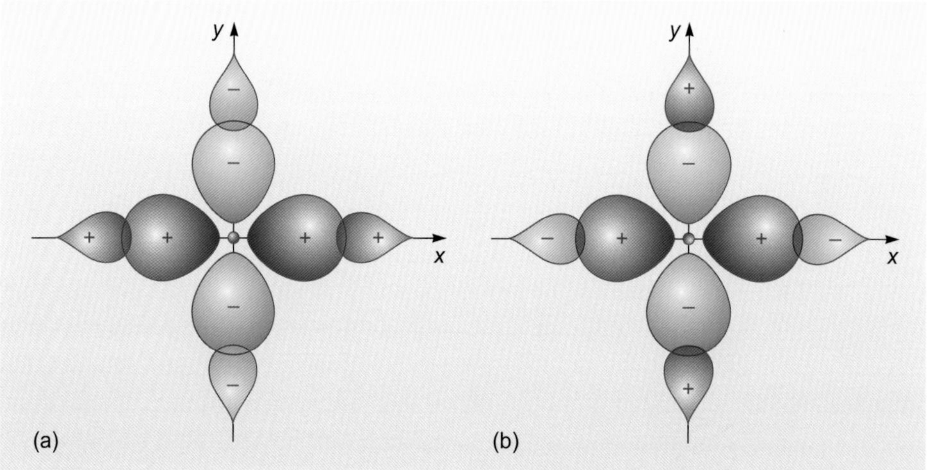

Figure 6.6 (a) A bonding combination of σ-bonding ligand orbitals on the x- and y-axes overlapping with a $d_{x^2-y^2}$ orbital on a metal atom, and (b) the antibonding combination of these orbitals.

From two metal d orbitals and six ligand orbitals, two bonding orbitals and two antibonding orbitals have been made. Can the remaining four combinations of ligand orbitals be combined with the other d orbitals? Do the σ-bonding ligand orbitals overlap with d_{xy}, d_{yz} or d_{xz}? Look at Figure 6.5b which shows a σ-bonding ligand orbital on the z-axis and a metal d_{xz} orbital.

■ Does the d_{xz} metal orbital form a bonding orbital with σ-bonding ligand orbitals on the *z*-axis?

□ No. Overlap with the positive lobe is cancelled by antibonding overlap with the negative lobe.

■ Does the d_{xz} metal orbital form a bonding orbital with σ-bonding orbitals on the *x*-axis?

□ No. As for the σ-bonding ligand orbitals on the *z*-axis, overlap with the negative lobe offsets that with the positive lobe.

A σ-bonding ligand orbital on the *y*-axis overlaps with all four lobes of d_{xz} but, as for the σ-bonding ligand orbital on the *z*-axis and the metal $d_{x^2-y^2}$ orbital, the *net* effect is non-bonding. So the metal d_{xz} orbital does not form any bonding orbitals for the complex with σ-bonding ligand orbitals. The d_{xy} and d_{yz} orbitals behave similarly. The three metal orbitals d_{xy}, d_{yz} and d_{xz} therefore remain **non-bonding** in the presence of σ-bonding ligand orbitals. As in crystal-field theory, these three orbitals are degenerate; they are labelled t_{2g}. In fact, the four remaining σ-bonding ligand orbitals combine with the metal 4s and 4p orbitals, forming four bonding and four antibonding orbitals.

6.1.1 Energy-level diagram for σ-bonded octahedral complex

Figure 6.7 shows a partial energy-level diagram for σ-bonded octahedral complexes. It comprises the metal 3d, 4s and 4p orbitals, the ligand σ-bonding orbitals and the complex orbitals. On the left are the metal orbitals, on the right, and at lower energy, are the ligand σ-bonding orbitals.

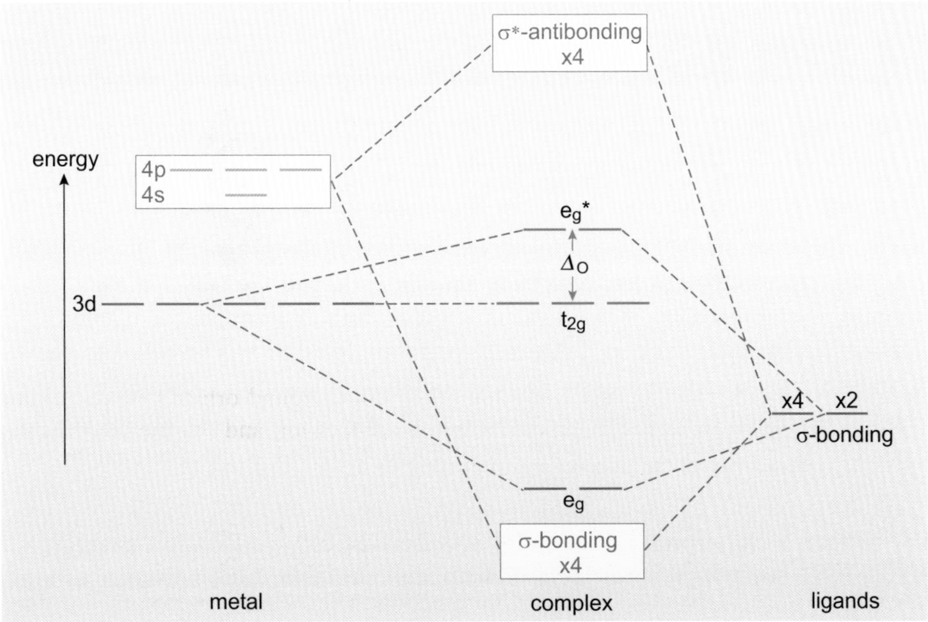

Figure 6.7 Partial energy-level diagram for a σ-bonded octahedral complex.

The lowest-energy complex orbitals are the four bonding orbitals formed from ligand orbitals and the 4s and 4p metal orbitals. Next are the e_g bonding orbitals. These are lower in energy than both the metal and ligand orbitals. The non-bonding metal d_{xy}, d_{yz} and d_{xz} orbitals form the t_{2g} orbitals at the same energy level as the metal d orbitals. Above both the metal d orbitals and the ligand orbitals, we have the antibonding e_g* orbitals. Finally we have the antibonding orbitals formed from ligand orbitals and the metal 4s and 4p. These lie above the 4s and 4p.

Strictly speaking, electrons from the ligands and the metal are indistinguishable, but for the purposes of filling in energy-level diagrams, we shall notionally indicate the electrons as being derived from the ligands or the metal.

The σ-bonding ligand orbitals contain two electrons each, so that there are 12 electrons from these orbitals to put into the orbitals of the complex. We can assign eight of these to the bonding combinations of ligand orbitals with 4s and 4p metal orbitals. Four of these electrons are assigned to the e_g bonding orbitals. This leaves the d electrons from the metal ion to fill the t_{2g} and e_g* orbitals.

Now let us compare this diagram with the crystal-field energy level diagram in Figure 5.5. The t_{2g} and e_g* orbitals play the role of the t_{2g} and e_g atomic orbitals in crystal-field theory. We can use these orbitals to explain properties just as we did with the t_{2g} and e_g atomic orbitals. The crystal-field splitting is replaced by the energy difference between the t_{2g} and e_g* levels, which is again labelled as Δ_o in Figure 6.7. These levels and the energy difference Δ_o between them correspond to those obtained by crystal-field theory, and are shown in green in Figure 6.7 and similar figures. To indicate that we are no longer confined to the simple crystal-field model, we shall from now on refer to the *ligand field* so that the energy difference between t_{2g} and e_g* becomes the **ligand-field splitting energy**. Using a reference state in which the t_{2g} and e_g* levels are equally occupied, we can work out the relative stabilities of transition-metal states, and thereby obtain a quantity equal to the crystal-field stabilisation energy, which we shall call the **ligand-field stabilisation energy, LFSE**. For example, a complex of a d^1 metal ion has one electron in t_{2g} and none in e_g*. We can define a reference state as one in which the electron occupies t_{2g} and e_g* equally, so that it spends $\frac{3}{5}$ of its time in t_{2g} and $\frac{2}{5}$ of its time in e_g*. The e_g* level is Δ_o higher in energy than the t_{2g}. Thus if we define the t_{2g} level as zero energy, a d electron in the reference state will have an energy of $\frac{3}{5} \times 0 + \frac{2}{5} \times \Delta_o = \frac{2}{5}\Delta_o$. By confining the electron to t_{2g} its energy will drop to 0. Thus the difference in the energy between an electron in t_{2g} and an electron in the reference state is $\frac{2}{5}\Delta_o$ and we say that the complex is stabilised by $\frac{2}{5}\Delta_o$. That is, the complex is more stable by this amount. If you look back at Table 5.2, you will see that this is equal to the crystal-field stabilisation energy for a d^1 octahedral complex. Similarly, if the electron occupied e_g*, its energy would be Δ_o and the complex would be stabilised by $(\frac{2}{5}\Delta_o - \Delta_o) = -\frac{3}{5}\Delta_o$. That is, the complex is destabilised by $\frac{3}{5}\Delta_o$.

■ What are the orbital occupancies and ligand-field stabilisation energy (LFSE) for a high-spin (weak-field) octahedral σ-bonded complex of a d^7 ion?

□ Five of the seven electrons occupy t_{2g} and two go into e_g*, giving a configuration $t_{2g}{}^5 e_g{}^{*2}$. Each electron in t_{2g} contributes a ligand-field stabilisation energy of $\frac{2}{5}\Delta_o$, and each electron in e_g* contributes $-\frac{3}{5}\Delta_o$. The total LFSE is thus $(5 \times \frac{2}{5}\Delta_o) + (2 \times -\frac{3}{5}\Delta_o) = \frac{4}{5}\Delta_o$, in agreement with the value in Table 5.2.

So far we have simply reproduced the crystal-field results but we can go further. As well as enabling us to work out LFSEs, diagrams such as Figure 6.7 give us some insight into what makes a particular ligand either strong-field or weak-field. The size of the energy gap, Δ_o, depends on the strength of the σ-bonding between metal and ligand; the stronger the bonding the greater is the energy gap and therefore the stronger the ligand field. To form a strong bond, the ligand must have a filled σ-bonding orbital close in energy to that of the metal d orbitals, and which overlaps well with the d orbitals.

6.1.2 Tetrahedral complexes

For tetrahedral complexes, the four ligands are arranged at the corners of a cube with the *x*-, *y*- and *z*-axes going through the cube faces (Figure 5.26) as in crystal-field theory. The σ-bonding ligand orbitals overlap in this case with the d_{xy}, d_{yz} and d_{xz} metal orbitals, shown, for example, for $3d_{xz}$ in Figure 6.8. In complexes of tetrahedral symmetry, as you saw in Section 5.6, these orbitals are labelled t_2. There is no overlap with the $d_{x^2-y^2}$ and d_{z^2} orbitals which remain non-bonding and are labelled e.

- ■ Why is there no g subscript on these labels?
- □ Tetrahedral complexes do not have a centre of symmetry. The subscript g refers to behaviour of the orbital when inverted through the centre of symmetry, and hence cannot be used for orbitals in tetrahedral complexes.

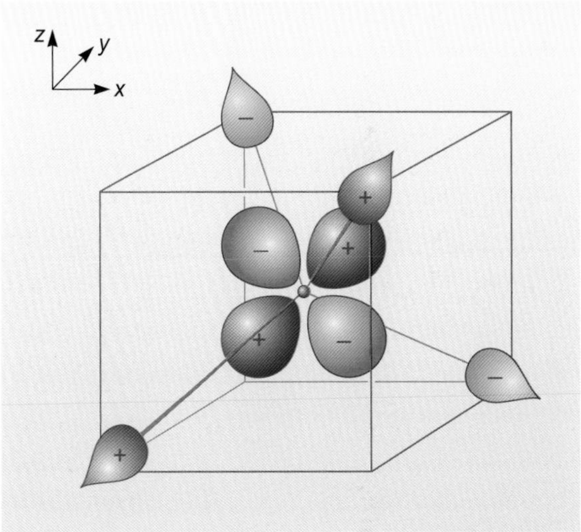

Figure 6.8 σ-bonding ligand orbitals forming a tetrahedron around a metal $3d_{xz}$ orbital.

Because the σ-bonding orbitals do not point directly at the d orbitals (see Figure 6.8), the σ-bonding in tetrahedral complexes is weaker than in octahedral complexes. Thus, the energy gap between the t_2 bonding and t_2* antibonding orbitals is less than that between the e_g bonding and e_g* antibonding orbitals in octahedral complexes (Figure 6.9). The fourth σ-bonding ligand orbital overlaps with the metal 4s.

Δ_t will be the energy gap between e and t_2* and due to the weaker bonding, this will be less than Δ_o.

6.2 π-bonding in strong-field complexes

σ-bonding by itself, however, is not enough to explain the spectrochemical series; the strongest-field ligands such as CO and PR_3 owe their strong metal–ligand bonds to their ability to form π bonds as well as σ bonds. It is to this π-bonding that we now turn. This enables us to begin to account for the somewhat unexpected ordering of ligands in the spectrochemical series described in Section 5.4.

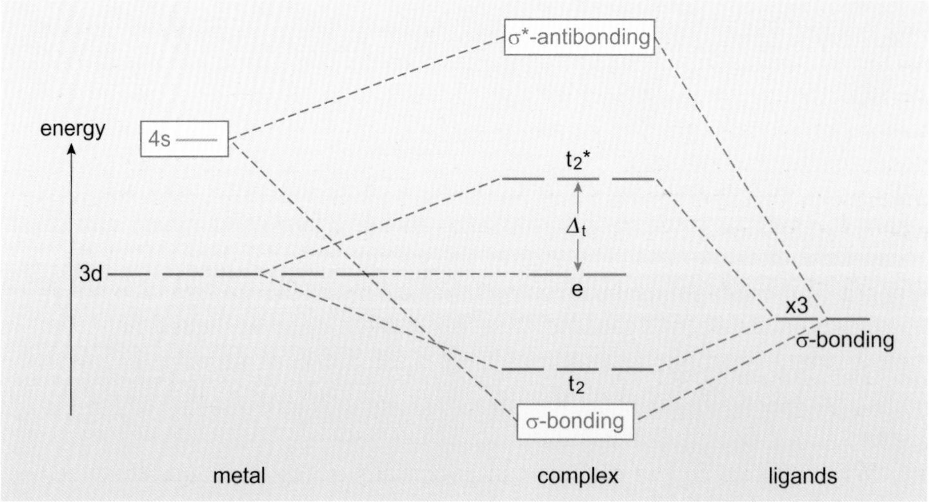

Figure 6.9 Partial energy-level diagram for a σ-bonded tetrahedral complex.

Many ligands such as water, ammonia and halide ions form complexes by donating a lone pair of electrons, thereby acting as σ donors to metal atoms or ions. These form stable complexes with both main-Group and transition elements. However, there are molecules which have no significant ability to form complexes with main-Group elements, but do form complexes with transition metals. These molecules, such as carbon monoxide, CO (carbonyl ligands); phosphines, PR_3 (phosphino ligands); nitric oxide, NO (nitrosyl ligands); the cyanide anion, CN^- (cyano ligands), form a wide range of compounds with transition elements using π orbitals to bond.

This type of bonding can be strong. The poisonous nature of CO and CN^- arises from the strong bonds they form to iron in haemoglobin and other important molecules in the body.

π-bonding ligand orbitals are antisymmetric to rotation about the metal–ligand bond; they form complex orbitals in which the electron density is concentrated above and below the bond (as in the π orbitals of diatomic molecules). Of particular importance for strong-field ligands are empty π-bonding orbitals whose energy is close to that of the metal d orbitals such as the 2pπ* in CO and CN^-. These orbitals are antibonding in CO and CN^- but we shall refer to them as **π-bonding orbitals** because they are involved in π bonds between the ligand and the metal.

First we will consider carbonyl complexes in some detail. These complexes can have metal–metal bonds, and the carbonyl ligands can be in two distinct positions – terminal with the carbon atom bonded to one metal atom, and μ_2-bridging with the carbon linked to two metal atoms. However, it is the nature of the bonding to the terminal CO that is most interesting.

6.2.1 π back-bonding

The CO–metal bonding can be described as comprising two parts: a weak σ bond using electrons in mainly CO orbitals which are donated *to* the metal, and a stronger π bond using electrons in mainly metal d orbitals which are donated *by* the metal to the ligand. In Chapter 3 you saw that these ligands are known as π-acceptor ligands. It is this donation of electrons from the metal to the empty π-bonding orbitals that gives rise to this description. Such complexes are not found in complexes of the s and p group metals as these metals do not have suitable d electrons for donation. Before looking at the molecular orbital picture in detail, it is useful to look at a pictorial model. You saw the molecular orbital diagram for CO in Figure 6.1.

■ Which CO orbital would take part in σ-bonding in an octahedral metal carbonyl complex, $M(CO)_6$?

□ The 5σ orbital (the HOMO), which contains two electrons and will overlap with vacant $d_{x^2-y^2}$ and d_{z^2} on the metal, to form a σ bond.

■ Which vacant ligand orbitals can potentially accept electrons from the metal d orbitals?

□ The 2π* orbital (the LUMO) is a potential candidate.

As before, it is important that the ligand orbital has the correct symmetry to overlap with the metal d orbitals. Considering still the octahedral complex, $M(CO)_6$, Figure 6.10 shows that the CO 2π* orbital has the correct symmetry to interact with a $3d_{xz}$ orbital on the metal.

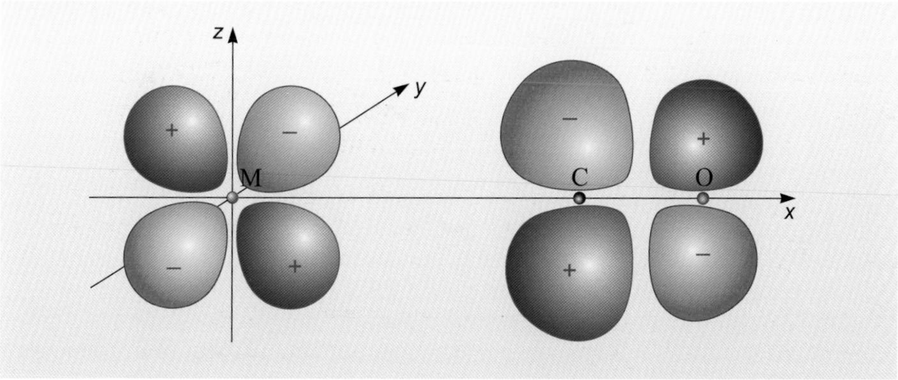

Figure 6.10 A metal d_{xz} orbital and an empty π*-antibonding molecular orbital on a CO ligand.

A π orbital is thus formed – this is occupied by electrons from the metal d orbitals. This metal–ligand bond is additional to the metal–ligand σ bond and as the donation is now from the metal to the ligand, this type of bonding is commonly known as **π back-bonding**.

The drift of electrons from ligand to metal (σ-bonding) tends to make the metal more negatively charged. At the same time electrons move from the metal to the ligand (π-bonding). Thus the π bond strengthens the σ bond and vice versa. This mutual strengthening of the bonding is called the **synergic effect** or **synergic bonding**.

We will now consider π-bonding in general. In Section 6.1 we defined a generalised σ-bonding ligand orbital. We can expand our discussion to other π back-bonding ligands by introducing a generalised π-bonding ligand orbital.

Figure 6.11 shows how the d_{xz} orbital can interact with two generalised π-bonding orbitals on the *x*-axis and two on the *z*-axis to form a π-bonding orbital for the complex. The d_{xy} orbital forms a similar bonding orbital with π-bonding ligand orbitals along the *x*- and *y*-axes, and the d_{yz} orbital forms one with π-bonding ligand orbitals on the *y*- and *z*-axes. Thus, there are three degenerate π-bonding orbitals formed by the metal d_{xy}, d_{yz} and d_{xz} orbitals of the complex.

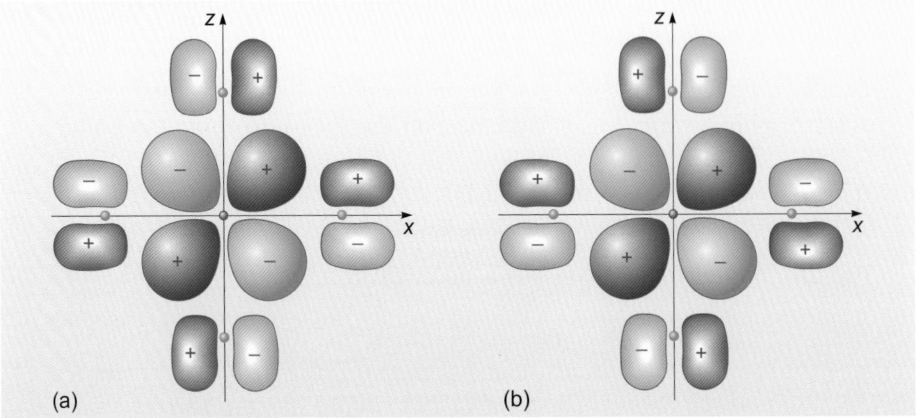

Figure 6.11 (a) π-bonding orbital for a metal complex formed from a metal d_{xz} orbital and generalised π-bonding ligand orbitals on the *x*- and *z*-axes, and (b) the corresponding antibonding orbital.

■ How should these orbitals be labelled?

□ t_{2g}. This label describes the symmetry of the d_{xy}, d_{yz} and d_{xz} orbitals of the metal in an octahedral complex, and so any molecular orbital formed from them will also have this label.

There will also of course be three degenerate antibonding π orbitals labelled t_{2g}*.

Do the d_{z^2} and $d_{x^2-y^2}$ metal orbitals form π bonds as well as σ bonds? The answer is no – Figure 6.12 shows a π-bonding ligand orbital on the *x*-axis and a metal $d_{x^2-y^2}$ orbital: the overlap with the positive lobe of the ligand π-bonding orbital is cancelled out by the overlap with the negative lobe. The net interaction between the $d_{x^2-y^2}$ orbital and π-bonding ligand orbitals on the other axes is also non-bonding, as is the interaction between the π-bonding ligands and the d_{z^2} orbital. Thus, as in crystal-field theory, the metal d orbitals in molecular orbital theory are divided into two sets – the d_{z^2} and $d_{x^2-y^2}$ orbitals, which overlap with σ-bonding ligand orbitals to form e_g orbitals on the complex, and the d_{xy}, d_{yz} and d_{xz} orbitals, which overlap with π-bonding ligand orbitals to form t_{2g} orbitals on the complex.

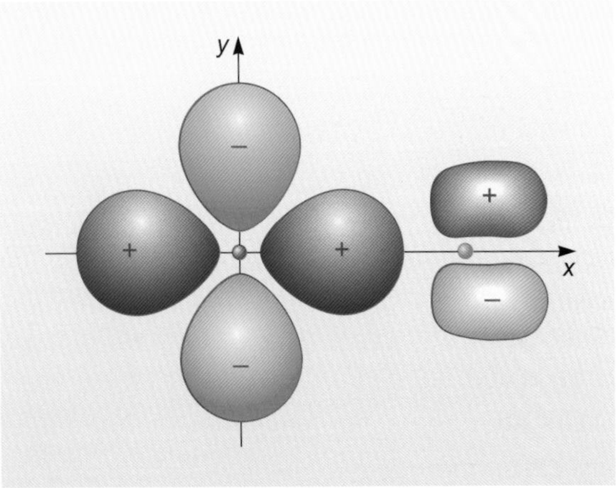

Figure 6.12 Overlap of a metal $d_{x^2-y^2}$ orbital and a π-bonding ligand orbital on the *x*-axis.

6.2.2 Energy-level diagram for carbonyl complexes

The energy-level diagram for an octahedral metal carbonyl complex, $M(CO)_6$ is shown in Figure 6.13, illustrating both σ- and π-bonding. On the left of the diagram are the d orbitals for the metal ion. On the right are the ligand orbitals. The ligand σ-bonding orbitals are full and of lower energy than the metal d orbitals. The ligand π-bonding orbitals are empty and of higher energy than the metal d orbitals.

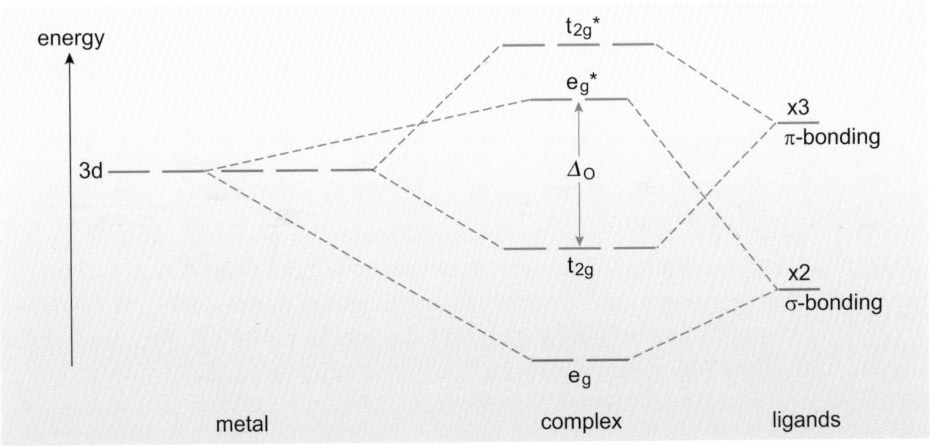

Figure 6.13 Partial energy-level diagram for an octahedral complex with both σ- and π-bonding.

There are six σ-bonding ligand orbitals; these form the two e_g and two e_g* orbitals (with d_{z^2} and $d_{x^2-y^2}$) leaving four orbitals that combine with 4s and 4p. Only the two orbitals that combine with d orbitals are shown in Figure 6.13.

CO has two doubly degenerate 2π* orbitals at right-angles to each other so for six CO ligands there are 12 orbitals altogether. With the three t_{2g} metal orbitals, these form the three t_{2g} and three t_{2g}* orbitals. This leaves nine ligand orbitals. These either combine with the 4s and 4p metal orbitals or are non-bonding. For clarity, these are again omitted from the diagram.

Because the ligand π-bonding orbitals are empty, there are only 12 electrons available from the ligands, two from each σ-bonding ligand orbital. Feeding these into the diagram, they fill the two bonding e_g orbitals and four bonding combinations of ligand orbitals with 4s and 4p metal orbitals. The metal d electrons can be thought of as feeding into the t_{2g} and e_g* levels. Δ_o is now the energy difference between t_{2g} and e_g*. In this diagram and subsequent orbital energy level diagrams, ligand orbitals that are non-bonding or overlap with metal 4s and 4p are omitted so that we can concentrate on the orbitals involving metal 3d orbitals.

The strong-field nature of CO can thus be explained by its ability to form strong π bonds with the t_{2g} metal orbitals, which leads to a lowering of energy of the t_{2g} orbitals of the complex and hence an increase of the energy gap, Δ_o. (Compare Figure 6.13 with Figure 6.7.)

6.2.3 Evidence for the nature of carbonyl bonding

X-ray diffraction

What is the evidence that this actually happens? We expect the C–O bond length to increase as it weakens, but unfortunately this is not a good indicator as it is rather insensitive to bond order and there is little variation in length between free carbon monoxide, C≡O bond (112.8 pm) and the lengths found for carbonyls (~115 pm). However, the strengthening (and shortening) of the metal–carbon bond can be detected from bond-length data, as M–C bond lengths tend to be sensitive to bond order. These bond lengths are obtained from X-ray crystal structures.

Infrared spectroscopy

Gaseous carbon monoxide, CO, vibrates at 2140 cm^{-1}. The additional electron density in a carbonyl antibonding system weakens the C–O bond and, consequently, the vibrational frequency is lowered from that of free carbon monoxide: the frequency range observed for terminal metal–carbonyl ligands is 1850–2125 cm^{-1}, and bridging carbonyl ligands in neutral molecules (which receive back-donation into the CO $2\pi^*$ orbital from two metals) show absorptions in the range 1700–1860 cm^{-1}.

6.3 Other π-acceptor complexes

Many other ligands have closely related bonding in transition-metal complexes to that described in Section 6.2 for carbonyl complexes: cyanide, nitric oxide, dinitrogen, phosphines, dioxygen, arsines, isonitrile, 2,2′-bipyridyl and others. Some are described below.

6.3.1 Cyanide ligands

The molecular orbital diagram for the CN^- ion is very similar to that for the isoelectronic CO molecule (Figure 6.1). Figure 6.14 illustrates how the electrons occupy the orbitals in Figure 6.13 for the complex $[Fe(CN)_6]^{3-}$, a strong-field octahedral complex of iron(III).

6.3.2 Nitrogen

Nitrogen is an essential element needed to make, for example, proteins, RNA, DNA and ATP. As nitrogen in the form of dinitrogen molecules makes up nearly 80% of our atmosphere, you might think that there would be no problem in obtaining sufficient nitrogen. However, dinitrogen is a very stable and kinetically inert molecule. The only living organisms that can use dinitrogen directly are the small proportion of bacteria that make the nitrogenase enzyme. No animals can fix nitrogen, nor plants; these rely on the products from nitrogen fixation to provide their nitrogen requirements. Even the bacteria are very inefficient at converting dinitrogen into a reduced form, ammonia (NH_3), which can be used to synthesise nitrogen-containing molecules.

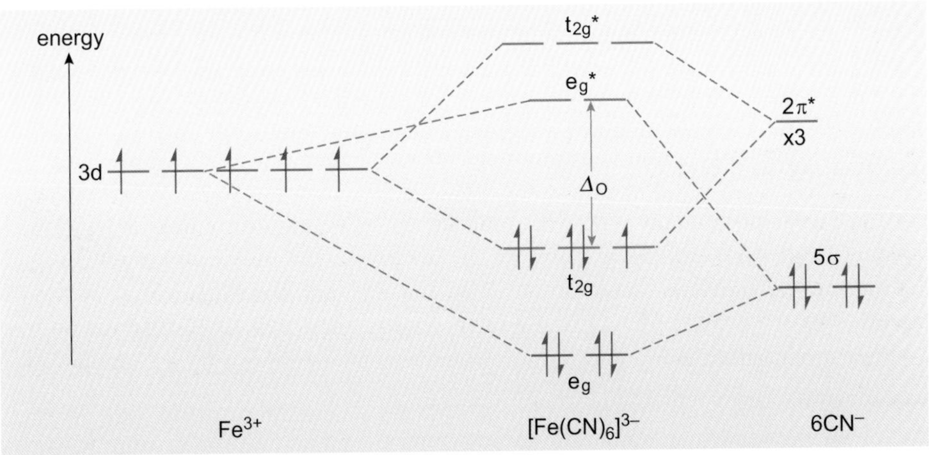

Figure 6.14 Partial energy-level diagram for $[Fe(CN)_6]^{3-}$. In energy-level diagrams for specific complexes, such as this, we only show for the ligands those orbitals that combine with metal d orbitals. This provides the correct number of electrons to feed into the orbitals of the complex.

In the 1960s, a desire to increase the production of ammonia stimulated research into dinitrogen complexes of transition metals and the first dinitrogen complex, $[Ru(NH_3)_5(N_2)]^{2+}$, was synthesised in 1965. Despite the inertness of dinitrogen, we now know that all transition elements can form dinitrogen complexes usually under mild conditions. However N_2 is less strongly bound than CO.

Dinitrogen is a π-acceptor ligand, isoelectronic with carbonyl, CO. The metal–ligand bonding is qualitatively the same as metal–CO bonding. However the binding is much weaker for N_2. Why should this be, given the similarities of the N_2 and CO molecules? The energies of the valence orbitals are similar for N_2 and CO, as might be expected, since nitrogen lies between carbon and oxygen in the Periodic Table. But the energy of the σ-donor orbital of N_2 is relatively low and the energy of the π* orbital relatively high, and so its orbitals are not as good a match for the d orbitals of most transition metals as are the corresponding orbitals of CO. As N_2 is a relatively poor σ-donor and a moderate π-acceptor, it is less strongly bound than CO. This is reflected in the lower values of Δ_o for $[M(N_2)_6]$ complexes than for $[M(CO)_6]$ complexes.

6.3.3 Nitric oxide

Nitric oxide, NO, is an important signalling molecule in the body, having roles as a muscle relaxant, vasodilator and neurotransmitter. Organic nitrates such as amyl nitrite, nitroglycerin and more recent drugs for angina, act to lower blood pressure by releasing NO. In many enzymes it coordinates to Fe.

NO has one electron more than CO and CN^- (in the $2\pi^*$ antibonding orbital, see Figure 6.1). The M–N–O system is commonly linear, but many complexes are known in which it is bent. The linear complexes can be considered as NO^+ complexes (NO^+ being isoelectronic with N_2) and the bent complexes as NO^- complexes (isoelectronic with O_2). Linear and bent forms can be distinguished by their NO stretching frequencies and ^{15}N NMR chemical shifts. For example, the linear nitrosyl complexes have NO stretching frequencies in the range 1700–1900 cm^{-1} close to the stretching frequency in NO, 1876 cm^{-1}. This is as expected for a bound NO^+ molecule. NO^+ has a bond order of 3 which would be reduced to nearer 2.5, as in NO, by coordinating to the metal.

For the linear complexes with NO binding as NO^+, as with CO and N_2, a σ bond is formed by donation from the 5σ orbital to the metal and electron density on the metal is 'back-donated' to the 2π*-antibonding orbital of the ligand. The π* orbital of NO is lower in energy than that of N_2 and so NO is a better π-acceptor.

NO^- will have two electrons in the 2π* orbital. We shall consider one NO ligand on the *z*-axis. A bent arrangement allows for a direct σ-overlap between the d_{z^2} orbital of the metal atom and the 2π* orbital of the NO which lies in the metal *xz*-plane (Figure 6.15a). This is in addition to the σ-overlap between the d_{z^2} orbital and the 5σ orbital. Thus we have overlap of three orbitals, 5σ, 2π*, and d_{z^2}. Combining these orbitals will give three orbitals for the complex. These are a σ-bonding orbital, a σ-antibonding orbital and a non-bonding orbital. The non-bonding orbital is concentrated on the N atom. The two 2π* electrons occupy this non-bonding orbital. The σ-bonding with the d_{z^2} orbital stabilises the complex. The other 2π* orbital of the NO forms a weak π bond with the d_{yz} orbital of the metal (Figure 6.15b). This bent M–N–O geometry is found when NO attaches to Fe in haem-containing enzymes.

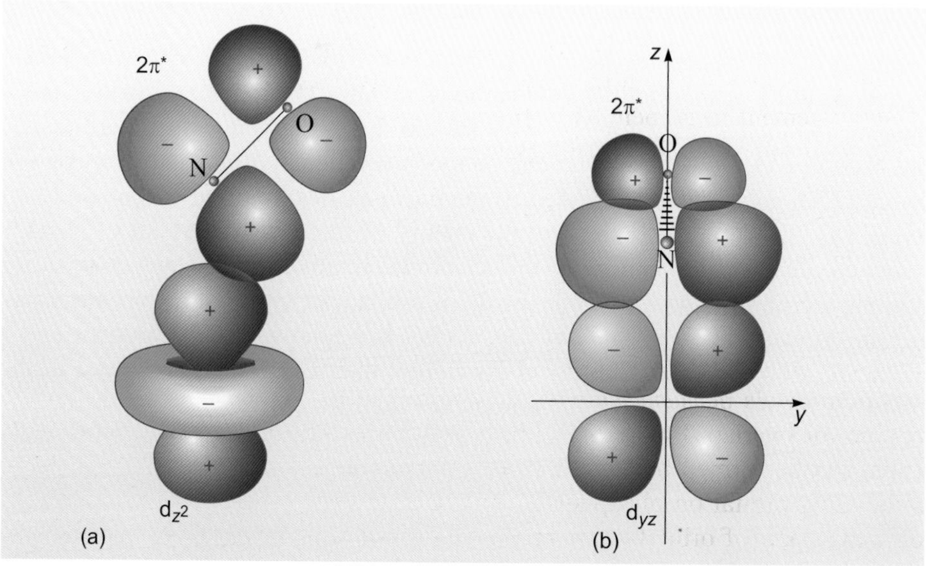

Figure 6.15 (a) Overlap of NO and transition-metal orbitals. The 2π* orbitals of NO form (a) a σ bond with the d_{z^2} orbital on the metal, and (b) a weak π bond with the metal d_{yz} orbital.

Many NO complexes are known, including $[Fe(NO)_2(CO)_2]$, $[Mn(NO)_3(CO)]$ and $[V(NO)(CO)_5]$, but there are few complexes in which nitrosyl is believed to be the unique ligand, $[Cr(NO)_4]$ being an example.

6.3.4 Oxygen

Dioxygen, O_2, binds reversibly to haemoglobin every time we breathe in, and interest in how it binds to and is released from haemoglobin has led to much research on O_2 complexes. Oxygen will react in a reversible manner with transition-metal complexes to give a variety of O_2 coordination modes (summarised in Figure 6.16).

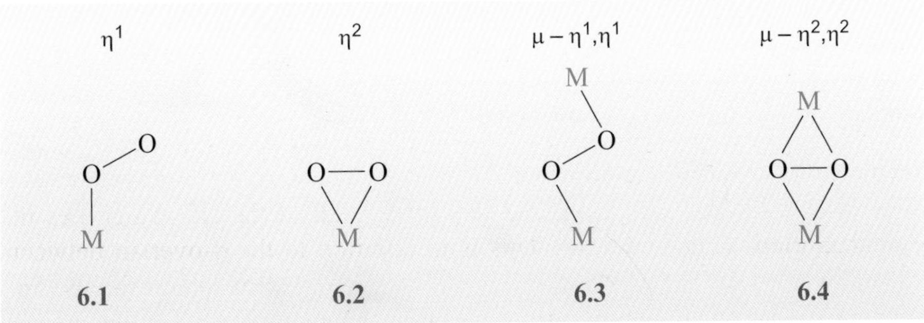

Figure 6.16 O_2 coordination modes with transition-metal complexes.

O_2 uses similar orbitals to CO, CN^-, N_2 and NO, and as with NO^- there are two electrons in the π^* orbital.

■ Would you expect O_2 bonding in η^1 fashion to coordinate to transition metals to give a bent or linear M–O–O arrangement?

□ You saw above that NO^-, which is isoelectronic with O_2, coordinates to give a bent M–N–O arrangement so we might expect the M–O–O arrangement to be bent as well. O_2 binds end-on to iron with an M–O–O angle of about 140° in haemoglobin. You should note however that it is not always clear which oxidation state the metal is in and O_2 could be binding as O_2^- or even O_2^{2-} rather than O_2.

6.3.5 Phosphines

Phosphine ligands, PX_3, form π-acceptor complexes with transition metals because there are P–X antibonding orbitals suitable for overlap. Figure 6.17 illustrates how a molecular orbital can be formed between a metal $3d_{xz}$ orbital and an antibonding orbital on phosphorus: the positive lobes on each combining pair of orbitals overlap, as do the negative lobes.

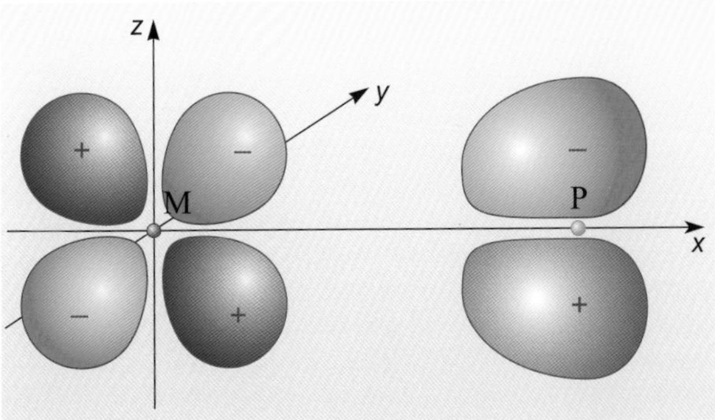

Figure 6.17 π-acceptor bonds between a metal d orbital and an empty antibonding orbital on phosphorus.

Carbon monoxide is displaced from many metal carbonyls by phosphine ligands to form such complexes as $[Cr(CO)_5(PPh_3)]$. The extent of the σ-donation from phosphorus and the back-bonding from the metal depends on the nature of the groups attached to phosphorus. For PH_3 or PR_3 (where R is an alkyl group), the π-acceptor ability is low, but this increases with the addition of electronegative substituents on phosphorus. Consequently, PF_3 is an excellent π-acceptor, and is comparable to carbon monoxide in its π-bonding capacity. The effect of electronegative substituents on phosphorus can be seen reflected in the stretching frequencies of the CO bond in the complexes in Table 6.1. As the electronegativity of the substituents increases, more electron density is donated from the metal to the phosphorus. Thus there is less back-donation to CO and the weakening of the CO bond is less. Thus with more electronegative substituents the CO bond is stronger and the vibrational frequency higher.

Table 6.1 Selected C–O stretching frequencies.

Complex	**σ(CO)/cm^{-1}**
$[Mo(CO)_3\{P(C_2H_5)_3\}_3]$	1937, 1841
$[Mo(CO)_3(PCl_3)_3]$	2040, 1991
$[Mo(CO)_3(PF_3)_3]$	2090, 2055

6.3.6 Ligands with extensive π bonds

You met both of these ligands earlier in Table 3.3 and in Section 5.4. They are repeated here for convenience.

The ligands 2,2′-bipyridyl (bipy; **6.5**) and 1,10-phenanthroline (phen; **6.6**) are capable of stabilising complexes with metals in low oxidation states through the involvement of the π system of the ligand. Ligands with extensive π systems are found widely in biological systems, for example haem in haemoglobin. These ligands coordinate to the metal via N or O but the orbitals on these atoms are linked to a delocalised π system in the rest of the molecule.

There are, however, some ligands for which the **delocalised π system** itself overlaps with the metal d orbitals; possibly the most famous such ligand is cyclopentadienyl, $C_5H_5^-$, **6.7**. As a simple example of the principles of this type of binding, we consider a simple alkene.

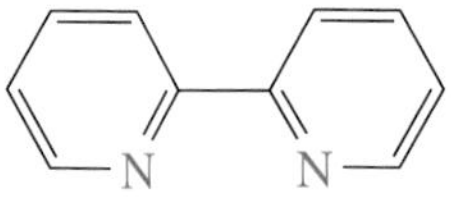

6.5

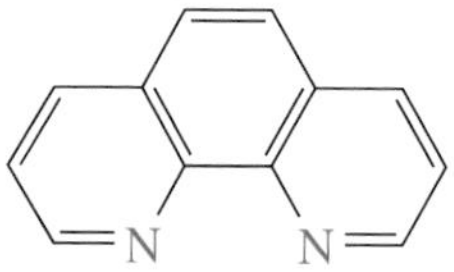

6.6

6.7

6.3.7 Alkenes

The bonding of ethene to transition metals is also closely related to that described for carbonyls and the other π-bonding ligands: in this case the σ bond is formed from the donation of π electrons on the alkene to empty metal orbitals, and there is also back-bonding from filled metal orbitals into the empty π* orbital of the alkene. These bonding principles are not unique to alkenes but can be applied to other systems such as allyls and alkynes and ligands with delocalised π systems such as cyclopentadiene. Included among these complexes are many catalysts of considerable industrial importance. A feature, which we look at in Section 6.4, of most of these complexes is that the transition metal occurs in a low oxidation state, often lower than the common oxidation states of the metals. Complexes of metals with low oxidation states are inherently thermodynamically unstable in the absence of π-acceptor ligands.

It is convenient to consider the σ-bonding component and the π-bonding component separately to begin with. In a simple example, Zeise's salt anion, $[PtCl_3(C_2H_4)]^-$, the ethene molecule bonds side-on to Pt (**6.8**).

The ethene π-molecular orbital is found to have suitable energy for overlap with the platinum 6s, 6p and 5d orbitals.

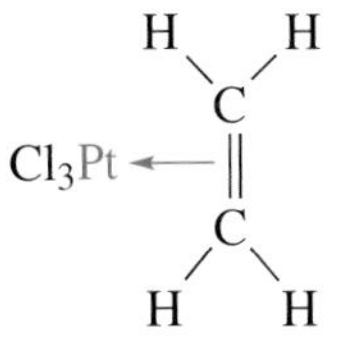

6.8

- ■ In separate diagrams, sketch the ethene π orbital together with the platinum orbitals that also have suitable symmetry for σ-bonding overlap with it. Assume that the platinum–ethene axis is the z-axis.

- □ Figure 6.18 shows that the platinum 6s, $6p_z$ and $5d_{z^2}$ orbitals have suitable symmetry (and energy) for overlap with the ethene π orbital. They can thus result in bonding.

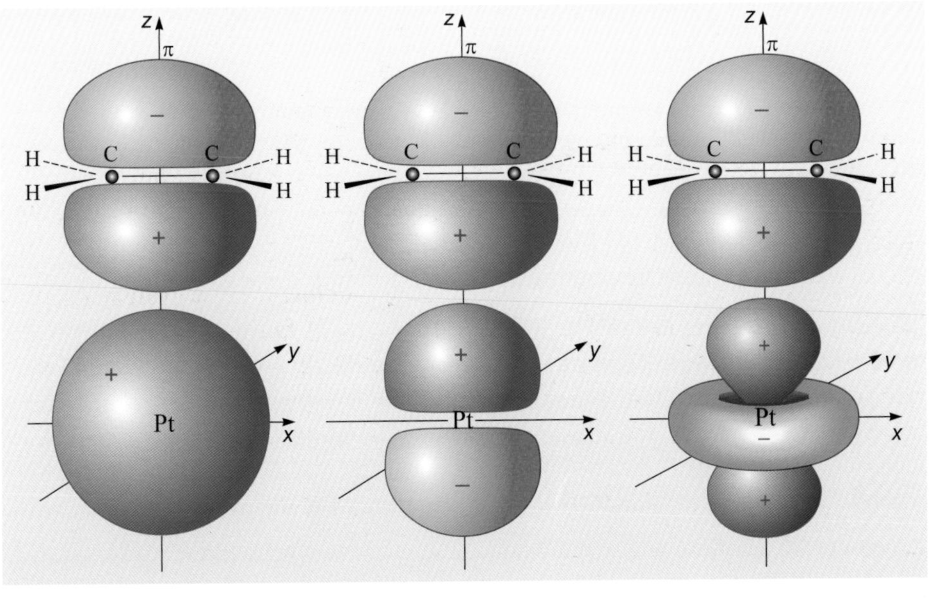

Figure 6.18 The platinum orbitals with the correct symmetry to overlap with the ethene π orbital to give σ-bonding.

Electron density is thus transferred from the filled π orbital on ethene to the empty orbitals on the metal. Metals towards the right of the Periodic Table have relatively less d-orbital involvement, because the d orbitals tend to become part of the core. If only metal–ligand σ-bonding were involved, most alkene complexes would not exist under normal conditions, because of the weak-donor properties of alkenes. Additionally such a process would cause a build-up of electron density on the metal that would tend to counteract the electron donation process, and so weaken the interaction. Back-donation from a filled metal orbital into an empty π* orbital of the alkene counteracts this.

- ■ Try to sketch this interaction for Zeise's salt anion, $[PtCl_3(C_2H_4)]^-$.

□ The platinum orbitals that have appropriate symmetry to overlap with the ethene π* orbital are those shown in Figure 6.19, the $6p_x$ and $5d_{xz}$.

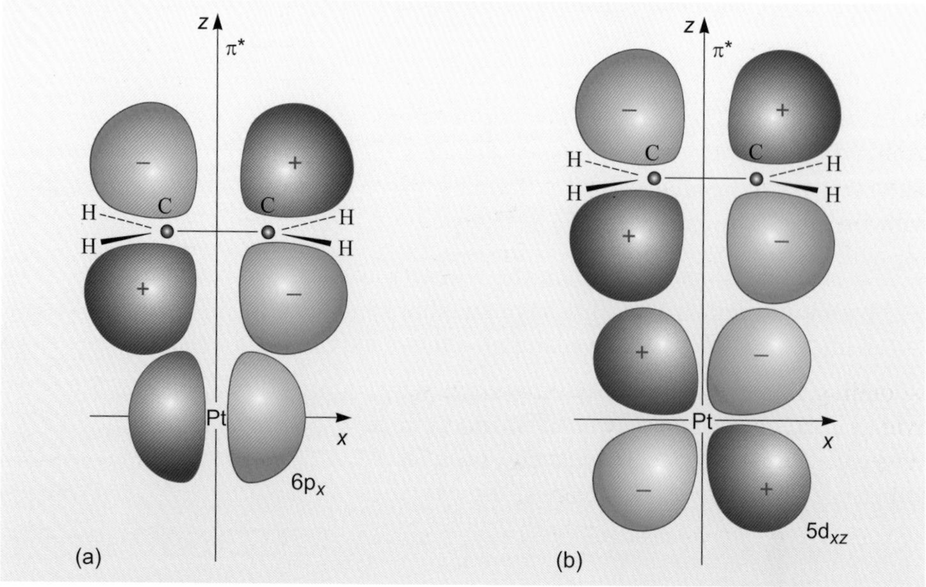

Figure 6.19 Bonding interactions between (a) the ethene π* orbital and the platinum $6p_x$ orbital, and (b) the ethene π* orbital and the platinum $5d_{xz}$ orbital for $[PtCl_3(C_2H_4)]^-$.

Metal electron density has effectively transferred back from platinum to ethene, and the process is often depicted by a simple line diagram as in **6.9**.

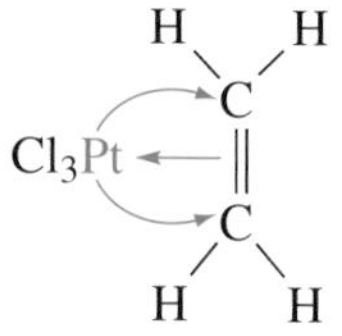

6.9

Overall there is donation of ethene π-electron density to platinum as a σ bond, and a reverse process of donation of metal electron density to the ethene π* molecular orbital. The two processes are not separable, but are synergic: the greater the σ-donation, the greater the π back-bonding from metal to alkene, because of the greater tendency to cause a charge build-up at the metal. This synergic process is similar to that which operates for metal carbonyls.

■ How would you expect the synergic process to affect the bond strength in ethene?

□ Electron density is *lowered* in the π-bonding orbital and *increased* in the π*-antibonding orbital, causing a lowering of the carbon–carbon bond order and a weakening of the bond.

In Zeise's salt anion, the consequent lengthening of the ethene bond is minimal (r(C=C) = 135 pm compared with 134 pm in ethene). However, in other complexes the consequences of the synergic process are often seen in the lowering of the ν(C=C) stretching frequency by 50–150 cm^{-1} and by the increase in the carbon–carbon distances of coordinated alkenes compared with free alkenes.

6.4 Oxidation states and the 18-electron rule

As we mentioned briefly above, low oxidation states in transition metals are stabilised by the back-donation of electron density from the metal d orbitals into suitable orbitals on the ligands as it moves the electron density away from the metal centre. It is useful at this point to look at oxidation states and an empirical rule for enabling the formulae of some complexes, particularly carbonyls, to be predicted. The rule is based on the premise (not always correct) that the total number of valence-shell electrons available to the metal is the same as that of the next noble gas; for the transition metals this will be 18, and for this reason the rule is known as the **18-electron rule**.

6.4.1 Oxidation states

π-bonding ligands stabilise low oxidation states of metals, that is the formal oxidation state of the metal in these complexes is low. For example, a typical range of compounds includes $[Fe(CO)_5]$, $[Mo(CO)_5I]^-$ and $[Ni(PMe_3)_4]$; as CO and PMe_3 are both neutral ligands, the oxidation state of the metal is 0 in each case. A few common carbonyl structures are shown in Figure 6.20.

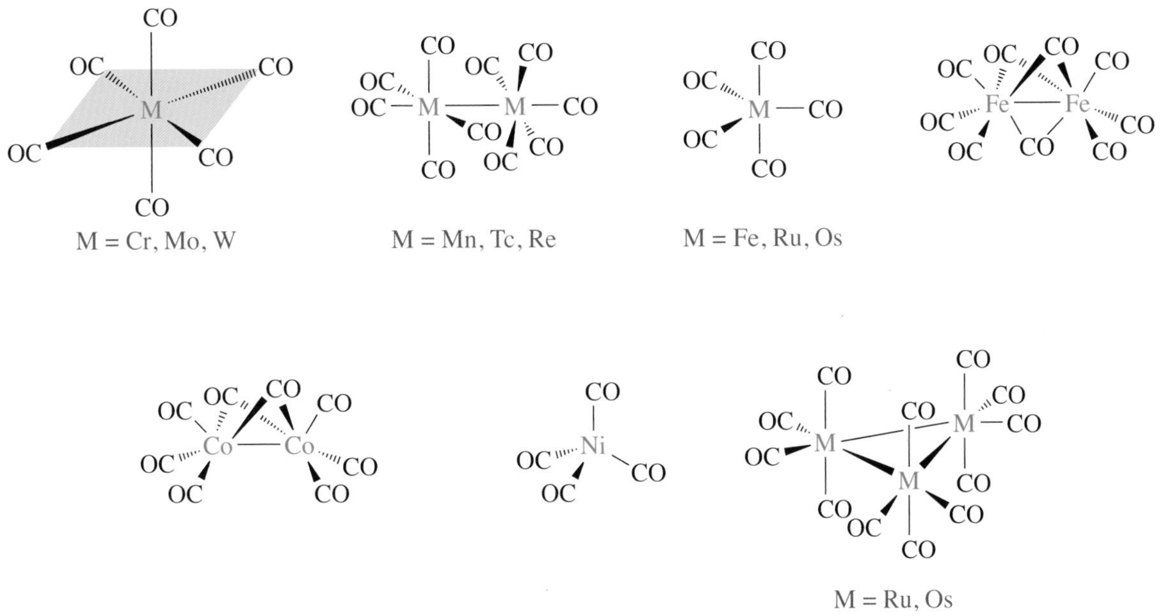

Figure 6.20 Structures of some neutral metal carbonyls.

In the sequence $[Mn(CO)_6]^+$, $[Cr(CO)_6]$ and $[V(CO)_6]^-$, the metals are in oxidation states +1, 0 and −1, respectively. The metals in carbonyl complexes are electron-rich relative to those in corresponding σ-bonded complexes such as $[Cr(H_2O)_6]^{3+}$.

6.4.2 The 18-electron rule

Consider a metal complex $[ML_6]$. The 18 electrons are accommodated in the nine valence-shell orbitals derived from the metal atom's five $(n-1)$d orbitals, three *n*p orbitals and one *n*s orbital (where *n* is the principal quantum number of the outer shell of electrons in the metal atom). Many classes of complex, particularly carbonyls, do follow the rule and so it can be used to predict the formulation and coordination number of the metal for individual complexes.

The total number of valence electrons in a complex comprises those donated by the ligands, together with the electrons associated with the metal. It is often impossible to assign meaningful oxidation states to metals and charges to ligands, particularly when extensive molecular orbitals are involved in their bonding. Thus for ease of electron counting, we adopt the convention here that the metal is in oxidation state 0, even when it is actually in some other state, and treat each ligand as a neutral molecule or as a radical. If the complex is charged, electrons are added or subtracted.

When the metal is considered to be in oxidation state 0, a terminal Cl atom (s^2p^5) donates *one* electron to a metal from its half-filled orbital, as do the other halogens and hydrogen; when chlorine acts as a bridging ligand, it uses one lone pair as well as its singly occupied orbital, so it is then a three-electron donor. CO and PR_3, when terminally bound, donate electrons via lone pairs, and are thus two-electron donors; a bridging carbonyl group donates a total of two electrons to all the metal atoms involved. Simple alkenes use two electrons for bonding to metal atoms; an alkyl group provides one electron. A metal–metal single bond donates one electron to each metal.

The rules are summarised in Table 6.2.

Table 6.2 Number of electrons donated from various ligands in different coordination modes for application of the 18-electron rule.

Ligands	κ^1	μ	η^2
CO	2	2	
NO	3		
F, Cl, Br, I, H	1	3	
alkyl group, R = methyl, ethyl, etc.	1	1	
ammonia, NH_3, NR_3, phosphine, PH_3, arsine, AsH_3	2	2	
ethene			2

We can illustrate this approach for $[MnCl(CO)_5]$; the manganese atom has ten electrons available from five CO ligands, one electron from chlorine and seven from manganese (d^5s^2 configuration), giving a total of 18 electrons.

■ Suggest the formulae of uncharged complexes that obey the 18-electron rule containing:

(a) carbon monoxide bonded to chromium; (b) CF_3 and/or CO groups bonded to iron.

□ (a) Cr ($4s^13d^5$) provides six electrons and the six CO ligands twelve, so $[Cr(CO)_6]$ is an 18-electron system: this complex is well known and stable.

(b) Iron ($4s^2 3d^6$) provides eight electrons, so to achieve a total of 18, ten additional electrons are required. These could be provided by five × two-electron donors (CO) to form $[Fe(CO)_5]$; or two × one-electron donors (CF_3) and four × two-electron donors (CO) to form $[Fe(CO)_4(CF_3)_2]$.

The 18-electron rule is followed by most transition-metal complexes with organic ligands, **organometallic complexes**, in which they are in a low oxidation state, although exceptions do occur towards the two extremes of the d-block in the Periodic Table. Metals with d^8 configurations (Pt(II), Pd(II), Ir(I), Au(III)) have a particularly strong tendency to form 16-electron square-planar complexes.

Cyano complexes often fail to obey the 18-electron rule, as in the series $[Cr(CN)_6]^{3-}$, $[Mn(CN)_6]^{3-}$ and $[Fe(CN)_6]^{3-}$ (15, 16 and 17 electrons, respectively). However, there do not appear to be any stable cyano complexes that *exceed* the 18-electron criterion: for example, on dissolving cobalt(II) cyanide in aqueous potassium cyanide, $[Co(CN)_5]^{3-}$ is obtained rather than $[Co(CN)_6]^{4-}$.

6.5 π-bonding in weak-field complexes

For some ligands such as halide ions, the effect of π-bonding is to *weaken* the ligand field. These are ligands that have *filled* π-bonding orbitals close to, but lower than, the metal d orbitals in energy.

■ Which orbitals on halide ligands will these be?

□ These orbitals are the two *n*p orbitals on each ion, which are not involved in σ-bonding.

A partial energy-level diagram showing only those complex orbitals that involve the metal d orbitals is shown in Figure 6.21. Ligand orbitals that are non-bonding or combine with metal 4s and 4p orbitals are omitted for clarity.

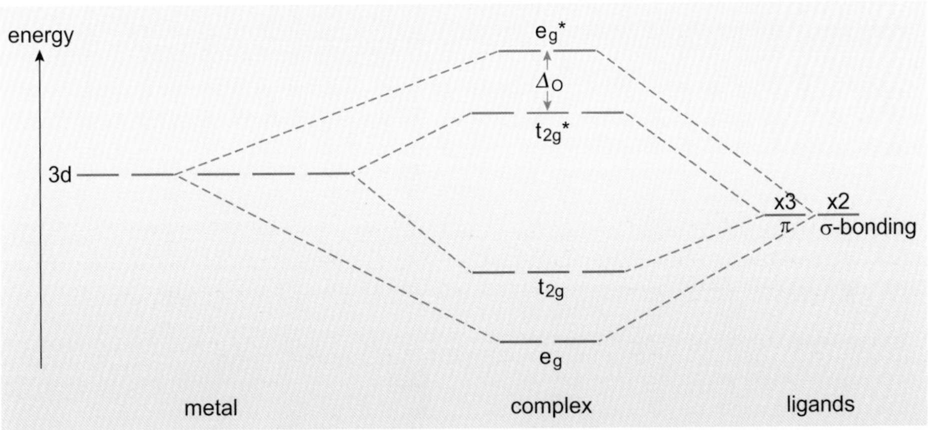

Figure 6.21 Partial energy-level diagram for an octahedral complex, in which filled σ- and π-bonding ligand (such as halide ions) orbitals interact with the metal d orbitals.

How do such orbitals weaken the ligand field? The filled p, π-bonding, orbitals overlap with the metal d_{xy}, d_{yz} and d_{xz} orbitals to form t_{2g} and t_{2g}^* orbitals. *These π-bonding orbitals are lower in energy than the metal d orbitals* (as are the σ-bonding p orbitals) and lie below the filled ligand p (π-bonding) orbitals.

For ions such as halide (X^-), oxide (O^{2-}) and nitride (N^{3-}), for example, both the σ- and π-bonding ligand orbitals are *n*p orbitals. Generally σ bonds are stronger than π bonds, and so where the σ- and π-bonding ligand orbitals are at the same energy, the σ-bonded e_g orbitals lie below the π-bonded t_{2g} orbitals. Consequently, the antibonding e_g^* orbitals lie *above* the antibonding t_{2g}^* orbitals.

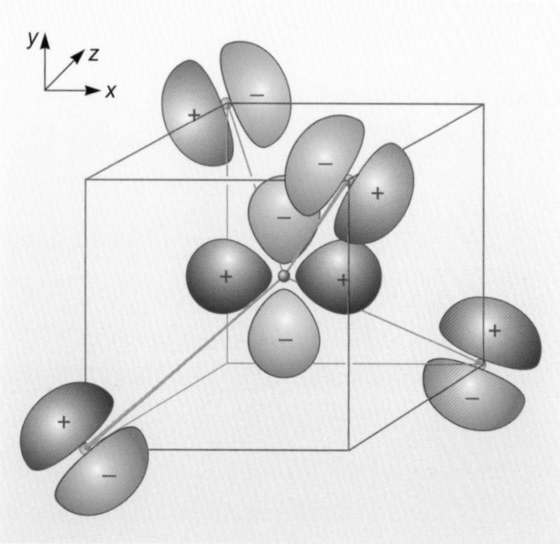

Figure 6.22 π-bonding ligand 2p orbitals overlapping with a metal $d_{x^2-y^2}$ orbital.

The electrons from the metal go into the t_{2g}^* and e_g^* orbitals and the ligand-field splitting Δ_o is now between the t_{2g}^* and e_g^* levels. The effect of the filled π-bonding orbitals has been *to replace the non-bonding t_{2g} level of the σ-bonded complex by an antibonding t_{2g}^* level* – look back to compare with Figure 6.7. Because the antibonding level is higher in energy, the gap between the orbitals is reduced and hence the ligand-field splitting is less.

■ What will be the electronic configuration for a weak-field d^4 complex, with ligands having filled the π-bonding orbitals?

□ $t_{2g}^{*3}e_g^{*1}$.

Weak-field tetrahedral metal halo-complexes are also common. Like σ-bonding ligands, π-bonding ligands (Figure 6.22) do not overlap strongly with the metal d orbitals in these complexes.

There are thus both weak σ and π metal–ligand bonds in these complexes. As with octahedral complexes, the bonding orbitals are full and the metal d electrons go into e* and t_2^*. The energy difference between these two is Δ_t and because the bonding is weak, this is smaller than Δ_o, Figure 6.23.

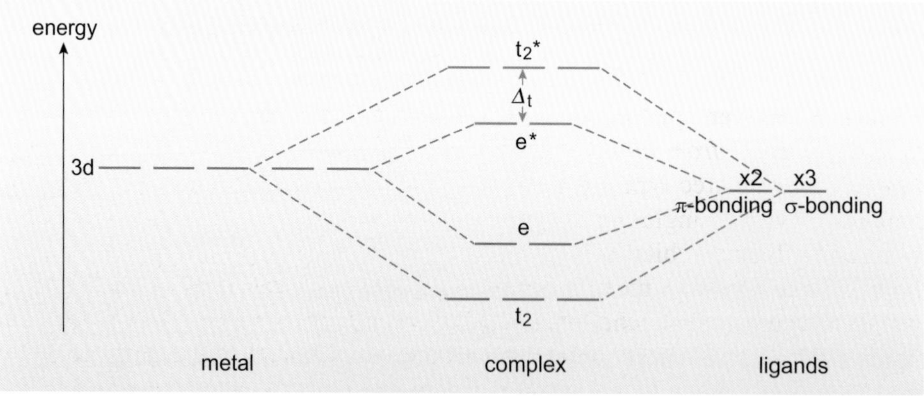

Figure 6.23 Partial orbital energy-level diagram for a tetrahedral complex formed from filled σ-bonding and π-bonding ligand orbitals.

6.6 Bonding and ligand field strength

We can now summarise the factors that affect the strength of the ligand field:

i Strong overlap between metal d orbitals and filled ligand σ-bonding orbitals close in energy leads to a strong field.

ii Overlap between metal d orbitals and *empty* π-bonding ligand orbitals of slightly higher energy strengthens the ligand field.

iii Overlap between metal d orbitals and *filled* π-bonding ligand orbitals lower in energy than the metal orbitals weakens the ligand field.

Thus, we would expect to find that a ligand with filled σ-bonding orbitals and empty π-bonding orbitals close in energy to the metal d orbitals would be a very strong-field ligand, and such is the case. Examples of ligands fulfilling these criteria are CO, CN^- and PR_3. The fluoride ion, on the other hand, with a filled π-bonding orbital, but no available empty π-bonding orbitals, is a very weak-field ligand.

Although we have only considered octahedral and tetrahedral complexes, similar considerations apply to other geometries.

6.7 Charge-transfer bands in the electronic spectra of transition-metal complexes

As you saw earlier, some of the deepest and most striking colours in transition-metal chemistry are not due to d↔d electronic transitions, but to so-called **charge-transfer transitions**. Examples include the yellow of the chromate(VI) ion (used for the pigments in roadside yellow lines), the orange of the dichromate(VI) ion, the deep purple of the manganate(VII) ion, the intense red of $[Fe(SCN)_4]^-$ and the orange of $TiBr_4$. The intensity of colour of some of these complexes, such as $[Fe(SCN)_4]^-$, accounts for their use in chemical analysis. The colours of the pigments yellow ochre (hydrated iron(III) oxide) and Prussian blue ($KFe(III)[Fe(II)(CN)_6]$) are also due to charge-transfer transitions. Such transitions even occur in biological systems such as the blue copper proteins (as you will see in Box 6.1).

But what are the characteristics of such transitions, and from which electronic transitions do they arise?

A **charge-transfer spectrum** arises because the absorbed radiation causes an electron to move *from an orbital based on one atom to an orbital based on a different atom*. Three types of charge transfer can be distinguished. For complexes with a single metal centre, transitions from a mainly metal orbital to a mainly ligand orbital or vice versa give rise to charge-transfer spectra. If there is more than one metal atom in the complex, metal-to-metal transitions can also occur, usually where the electron goes from a metal atom in one oxidation state to a metal atom in another oxidation state (for example, iron(II) to iron(III)). Charge-transfer spectra have two main characteristics:

- Charge-transfer transitions are generally very intense; their molar absorption coefficients are of the order of 10^3–10^4 dm^3 mol^{-1} cm^{-1}, in contrast to those of d–d transitions whose molar absorption coefficients are orders of magnitude less.

- The centre of the peak of a charge-transfer transition lies at a shorter wavelength (higher energy) than that of a d–d transition of the same complex.

In Section 5.4, you saw that d–d transitions were weak because they disobeyed the selection rules for electronic transitions.

■ What is the selection rule that d↔d transitions break?

□ The Laporte selection rule, which states that the orbital quantum number, *l*, can only change by ±1. Some very weak transitions also break the spin selection rule.

The spin selection rule still applies to charge-transfer transitions, but the Laporte selection rule only applies if the electronic transition is between two metal-based orbitals. If one of the orbitals is based on the ligands, it will not have a metal orbital quantum number and different rules come into play. The transitions that give rise to the intense absorptions are allowed transitions.

6.7.1 Ligand-to-metal transitions

In ligand-to-metal transitions, an electron jumps from an orbital of the complex mainly composed of ligand orbitals at low energy to one of the complex orbitals in which there is a large contribution from the metal d orbitals.

Interesting examples are the tetrahedral halo-complexes. For such complexes, the ligand orbital is a halogen *n*s or a combination of *n*s and *n*p orbitals. The promotion of electrons from halogen orbitals to the metal d orbitals gives rise to the charge-transfer spectra, for example, the orange colour of $TiBr_4$.

■ In titanium(IV) complexes how many electrons are there in the metal d orbitals?

□ There are no d electrons in titanium(IV) complexes.

Figure 6.24 is an orbital energy-level diagram for TiX_4 which, as well as the orbitals shown in Figure 6.23, includes some of the ligand orbitals involved in charge-transfer transitions.

The lowest-energy transition ($h\nu_1$) shown in Figure 6.24b is from a TiX_4 ligand orbital to the e* orbital. The next highest-energy transition ($h\nu_2$) is either from the ligand orbital to the t_2* orbital (Figure 6.24c), or from the bonding t_2 orbital (which is more ligand than metal in character) to the e* orbital (e↔e* is forbidden).

■ What is the difference in energy between the two transitions starting from the non-bonding ligand orbital?

□ The energy difference between e* and t_2* is the ligand-field splitting energy, Δ_t.

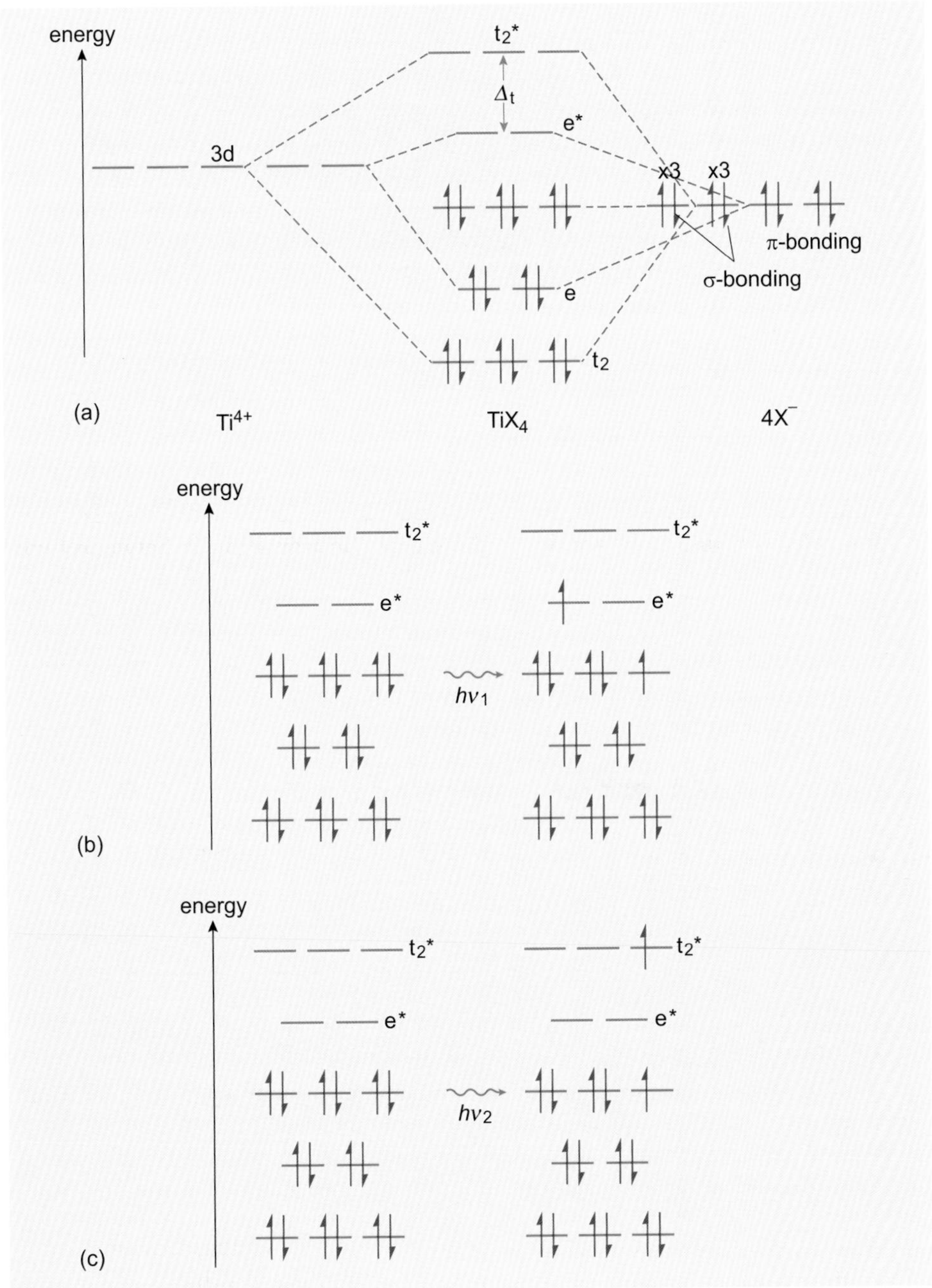

Figure 6.24 (a) Partial orbital energy-level diagram for TiX_4; (b) the lowest-energy charge-transfer transition; (c) one possibility for the next highest-energy charge-transfer transitions.

Thus, in some cases we can obtain ligand-field splitting energies from charge-transfer spectra.

The wavelengths observed in these spectra depend on the ease with which an electron can be removed from the ligand. The wavenumber of a particular ligand-to-metal charge-transfer band in halide complexes increases with an increase in the ionisation energy of the ligand *n*p electrons. This is because as the ionisation energy increases, the energy of the *n*p levels decreases, and so the energy gap between the ligand and metal orbitals increases. The ionisation energies increase in the order I < Br < Cl, so the wavenumber (and therefore energy) of the lowest energy transition in TiX_4 also increases in this order. The transition is at 19 600 cm^{-1} for TiI_4, at 29 500 cm^{-1} for $TiBr_4$ and at 35 400 cm^{-1} for $TiCl_4$. The two bands at 29 500 cm^{-1} and 35 400 cm^{-1} are in the ultraviolet region of the electromagnetic spectrum, but like many charge-transfer bands, the spectral bands corresponding to these transitions are very broad. It is the spread of the absorption band into the blue/violet region that gives $TiBr_4$ its orange colour.

For most tetrahedral transition-metal complexes, the d orbitals are partly occupied and it becomes more difficult to obtain the ligand-field splitting energy, as there is also an interaction of the promoted electron (e* or t_2*) with the electrons already occupying the mainly metal d orbitals. None the less, the shift in wavenumber of the charge-transfer spectral bands with the ionisation energy of the ligand is still clearly seen, as exemplified by the nickel(II) halide complexes in Figure 6.25.

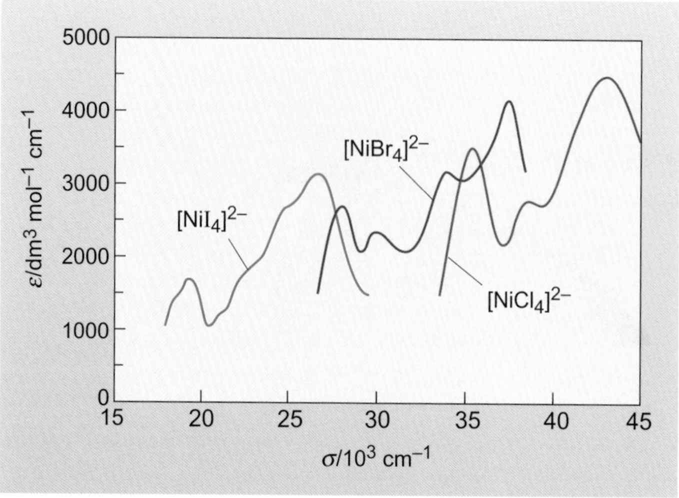

Figure 6.25 Charge-transfer bands in the electronic absorption spectra of the complexes $[NiX_4]^{2-}$.

Box 6.1 Blue copper proteins

Charge-transfer transitions can give rise to colour in molecules of biological origin. An example of biological molecules whose colour arises from a ligand-to-metal charge-transfer transition is a class of proteins known as *blue copper proteins*. Although blue is the colour typically found for copper(II) compounds such as hydrated copper sulfate, the blue copper proteins are a much deeper blue, and the transition giving rise to the colour has a molar absorption coefficient too large to be that of a d↔d transition. Azurin, plastocyanin (Figure 6.26) and stellacyanin contain copper ions surrounded by a distorted tetrahedron of two sulfur atoms and two nitrogen atoms, which originate from amino acid residues on the surrounding protein. Strong spectral bands arising from a transition from π-bonding orbitals on sulfur in a cysteine residue to the empty t_2* on copper occur at wavenumbers in the range 12 000–21 000 cm^{-1} (infrared to green).

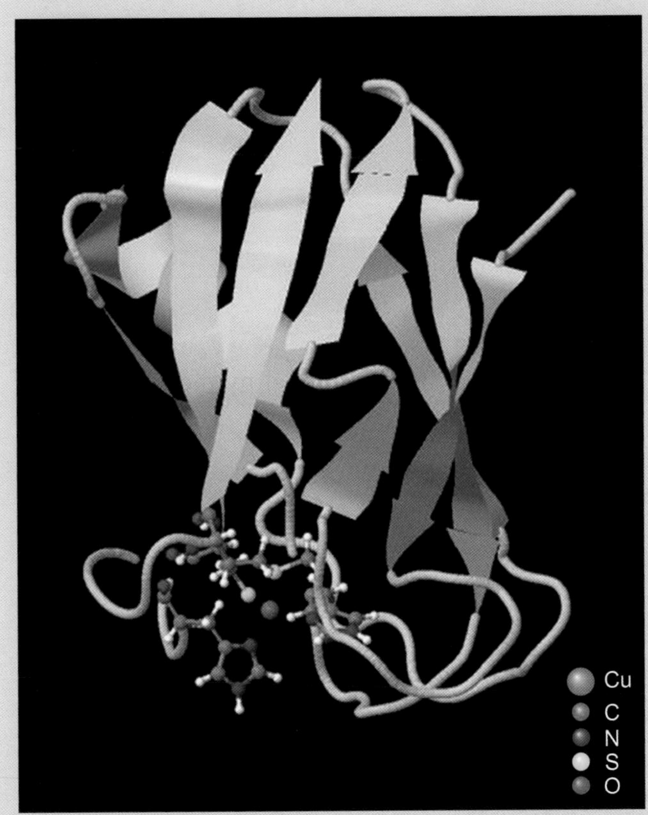

Figure 6.26 Schematic structure of plastocyanin. (Based on pdb file 1jxd (Bertini et al. 2001).)

6.7.2 Metal-to-ligand charge-transfer bands

If a complex has electrons in the metal d orbitals, they can be excited to unoccupied ligand orbitals lying at higher energy. Such a transition is a metal-to-ligand charge-transfer transition as charge (the electron) has gone from a metal-based orbital to a ligand-based orbital.

■ Which ligands give complexes with unoccupied orbitals lying not far above the metal d orbitals?

□ Those with unoccupied π orbitals at slightly higher energy than the metal d orbitals.

Charge-transfer bands of this type are thus characteristic of complexes containing strong-field ligands such as CO, CN^- and PR_3. Ligands with delocalised π orbitals, such as phen and bipy, also provide suitable empty π* orbitals. The intense red colour of bipy complexes of iron(II), for example, are due to a metal-to-ligand charge-transfer transition. For octahedral hexacyano complexes, the transitions are between the ligand-field orbitals, t_{2g} and e_g*, and the combinations of the unoccupied ligand π-bonding orbitals that do not overlap with the metal d orbital. The partial orbital energy-level diagram for $[Fe(CN)_6]^{4-}$ is shown in Figure 6.27 – this is the same as the one developed earlier in Figure 6.14 but now with some additional ligand orbitals. Absorption bands, found at 45 780 cm^{-1} ($h\nu_1$) and 50 000 cm^{-1} ($h\nu_2$), have been assigned to the transitions shown (Figure 6.27).

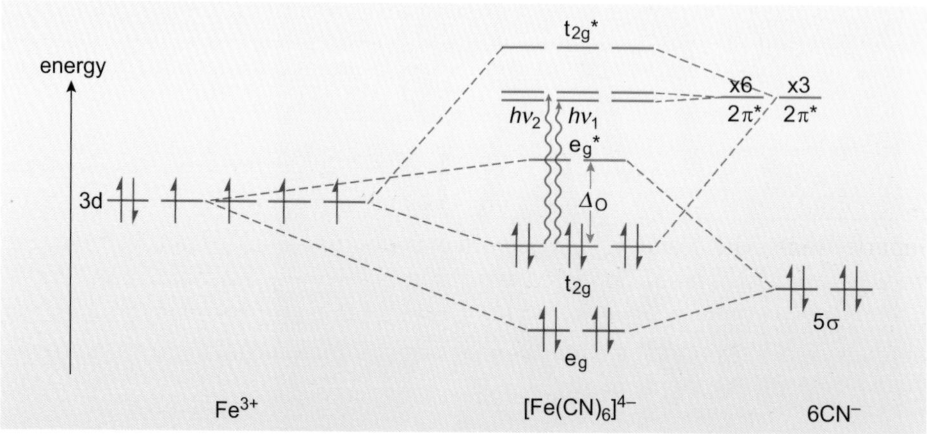

Figure 6.27 Partial orbital energy-level diagram for $[Fe(CN)_6]^{4-}$, showing the ligand orbitals involved in metal-to-ligand charge-transfer transitions.

6.7.3 Metal-to-metal transition

When there are two or more metal centres close together, spectral transitions can occur between the orbitals based on one metal and those on the other; in this case d↔d transitions are allowed. An interesting set of complexes is those with one metal in two different oxidation states. Such complexes are often intensely coloured. An example is the deep-blue compound known as Prussian blue, $KFe(III)[Fe(II)(CN)_6]$, in which the iron(III) ions are octahedrally surrounded by the nitrogen atoms of the $[Fe(CN)_6]^{4-}$ ions (Figure 6.28).

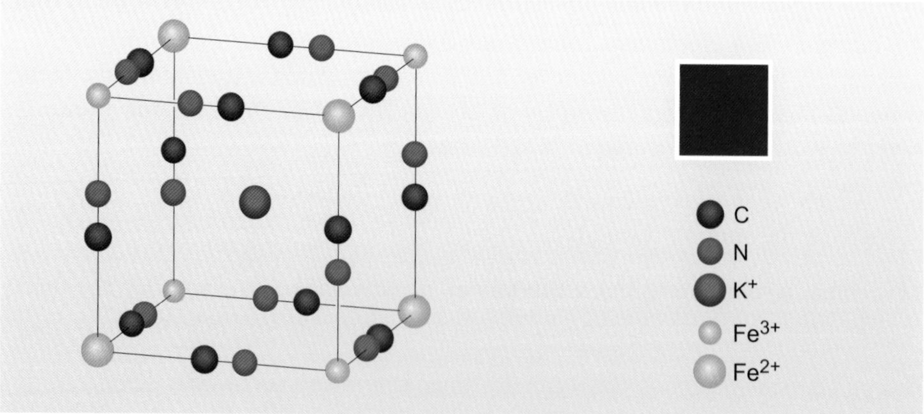

Figure 6.28 Part of a unit cell of Prussian blue, $KFe(III)[Fe(II)(CN)_6]$. The square inset shows the deep-blue colour of the compound.

In this complex the Fe(III) and Fe(II) atoms essentially act independently of each other.

The blue colour is due to transitions from a t_{2g} orbital on iron(II) in the $[Fe(CN)_6]^{4-}$ ion to the t_{2g} and e_g* orbitals on iron(III).

■ Why are the transitions only from a t_{2g} orbital of $[Fe(CN)_6]^{4-}$, and not from an e_g* orbital?

□ This complex will be strong field and iron(II) is d^6. In strong-field octahedral d^6 complexes, the electrons completely fill the t_{2g} orbitals, and the e_g* orbitals are empty.

Figure 6.29 A sapphire.

An example of a transition between two different metal centres is that giving rise to the colour of the gemstone sapphire. Like rubies, sapphires are based on corundum, but containing impurities. However, in this case, there are two impurity ions – Fe^{2+} and Ti^{4+}. The colour is due to a charge-transfer transition from Fe^{2+} to Ti^{4+} – a metal–metal transition between two different metals. This transition absorbs light in the red–yellow region producing the beautiful blue colour of sapphires, Figure 6.29.

Don't forget that there are questions on the companion website which you can use to test your understanding of the material covered in this chapter.

References

Bertini, I., Bryant, D.A., Ciurli, S., Dikiy, A., Fernández, C.O., Luchinat, C., Safarov, N., Vila, A.J., Zhao, J. (2001) 'Backbone dynamics of plastocyanin in both oxidation states. Solution structure of the reduced form and comparison with the oxidized state', *Journal of Biological Chemistry*, vol. 276, pp. 47217–26.

Mitić, N., Clay, M.D., Saleh, L., Bollinger, J.M. and Solomon, E.I. (2007), 'Spectroscopic and electronic structure studies of intermediate X in ribonucleotide reductase R2 and two variants: a description of the Fe(IV)-oxo bond in the Fe(III)–O–Fe(IV) dimer', *Journal of the American Chemical Society*, vol. **129**, issue 29, pp. 9049–65.

Acknowledgements

Grateful acknowledgement is made to the following sources for permission to reproduce material in this text.

Cover and title page

Sapphire gemstones. Lawrence Lawry / Science Photo Library.

Figures

Figure 5.17: Collection of Professor H Bank, Idar-Oberstein, Germany, courtesy of the Mineralogical Museum; Figure 6.29: Corundum, Laach lake volcanic complex, Collection Bernd Ternes, Photo Stephan Wolfsried. FOV 5 mm.

Every effort has been made to contact copyright holders. If any have been inadvertently overlooked the publishers will be pleased to make the necessary arrangements at the first opportunity.

Appendix

Standard redox potentials appropriate for acid solutions at 298.15 K

Electrode reaction	$E^{\ominus}$**/V**
$Li^{+}(aq) + e = Li(s)$	−3.04
$K^{+}(aq) + e = K(s)$	−2.94
$Rb^{+}(aq) + e = Rb(s)$	−2.94
$Cs^{+}(aq) + e = Cs(s)$	−2.92
$Ba^{2+}(aq) + 2e = Ba(s)$	−2.91
$Sr^{2+}(aq) + 2e = Sr(s)$	−2.90
$Ca^{2+}(aq) + 2e = Ca(s)$	−2.87
$Na^{+}(aq) + e = Na(s)$	−2.71
$Mg^{2+}(aq) + 2e = Mg(s)$	−2.36
$Al^{3+}(aq) + 3e = Al(s)$	−1.68
$Mn^{2+}(aq) + 2e = Mn(s)$	−1.18
$V^{2+}(aq) + 2e = V(s)$	−1.18
$Zn^{2+}(aq) + 2e = Zn(s)$	−0.76
$Cr^{3+}(aq) + 3e = Cr(s)$	−0.74
$Fe^{2+}(aq) + 2e = Fe(s)$	−0.46
$Cr^{3+}(aq) + e = Cr^{2+}(aq)$	−0.42
$Co^{2+}(aq) + 2e = Co(s)$	−0.28
$V^{3+}(aq) + e = V^{2+}(aq)$	−0.25
$Ni^{2+}(aq) + 2e = Ni(s)$	−0.24
$Sn^{2+}(aq) + 2e = Sn(s)$	−0.14
$Pb^{2+}(aq) + 2e = Pb(s)$	−0.13
$H^{+}(aq) + e = \frac{1}{2}H_2(g)$	0.00
$[Co(NH_3)_6]^{3+}(aq) + e = [Co(NH_3)_6]_3)^{2+}(aq)$	0.10
$Sn^{4+}(aq) + 2e = Sn^{2+}(aq)$	0.15
$Cu^{2+}(aq) + e = Cu^{+}(aq)$	0.16
$Cu^{2+}(aq) + 2e = Cu(s)$	0.34
$VO^{2+}(aq) + 2H^{+}(aq) + e = V^{3+}(aq) + H_2O(l)$	0.34
$[Fe(CN)_6]^{3-}(aq) + e = [Fe(CN)_6]^{4-}(aq)$	0.36
$\frac{1}{2}I_2(s) + e = I^{-}(aq)$	0.53
$MnO_4^{-}(aq) + e = MnO_4^{2-}(aq)$	0.56
$O_2(g) + 2H^{+}(aq) + 2e = H_2O_2(aq)$	0.68
$Fe^{3+}(aq) + e = Fe^{2+}(aq)$	0.77
$Ag^{+}(aq) + e = Ag(s)$	0.80
$VO_2^{+}(aq) + 2H^{+}(aq) + e = VO^{2+}(aq) + H_2O(l)$	1.00

$\frac{1}{2}Br_2(l) + e = Br^-(aq)$	1.08
$IO_3^-(aq) + 6H^+(aq) + 5e = \frac{1}{2}I_2(s) + 3H_2O(l)$	1.21
$O_2(g) + 4H^+(aq) + 4e = 2H_2O(l)$	1.23
$MnO_2(s) + 4H^+(aq) + 2e = Mn^{2+}(aq) + 2H_2O(l)$	1.23
$Tl^{3+}(aq) + 2e = Tl^+(aq)$	1.25
$Cr_2O_7^{2-}(aq) + 14H^+(aq) + 6e = 2Cr^{3+}(aq) + 7H_2O(l)$	1.36
$\frac{1}{2}Cl_2(g) + e = Cl^-(aq)$	1.36
$MnO_4^-(aq) + 8H^+(aq) + 5e = Mn^{2+}(aq) + 4H_2O(l)$	1.51
$BrO_3^-(aq) + 6H^+(aq) + 5e = \frac{1}{2}Br_2(l) + 3H_2O(l)$	1.51
$Mn^{3+}(aq) + e = Mn^{2+}(aq)$	1.60
$Ce^{4+}(aq) + e = Ce^{3+}(aq)$	1.61
$MnO_4^-(aq) + 4H^+(aq) + 3e = MnO_2(s) + 2H_2O(l)$	1.70
$Co^{3+}(aq) + e = Co^{2+}(aq)$	1.94
$S_2O_8^{2-}(aq) + 2e = 2SO_4^{2-}(aq)$	1.94
$Ag^{2+}(aq) + e = Ag^+(aq)$	1.98
$O_3(g) + 2H^+(aq) + 2e = O_2(g) + H_2O(l)$	2.07
$FeO_4^{2-}(aq) + 8H^+(aq) + 3e = Fe^{3+}(aq) + 4H_2O(l)$	2.20
$\frac{1}{2}F_2(g) + e = F^-(aq)$	2.89

Index

Entries and page numbers in **bold type** refer to key words which are printed in **bold** in the text. Indexed information on pages indicated by italics is carried mainly or wholly in a figure or a table.

X

Z

Periodic Table DVD-ROM information

The *Periodic Table* DVD contains:

- multimedia applications
- video sequences.

The disk is designed for use in a personal computer (PC). The DVD cannot be operated in a domestic DVD player connected to a television.

Computer specification

The DVD-ROM is designed for use on a PC running Windows XP or Vista. We recommend the following as the minimum hardware specification:

Processor 2GHz

Memory (RAM) 256 MB (512+ MB for Vista)

Hard disk space 50 MB

DVD drive

Video resolution 800 x 600 pixels at High Colour

Sound card and speakers/headphones

Computers with higher specification components will provide a smoother presentation of the multimedia materials

Installing the DVD software onto your computer

A few supporting software applications must be installed before you can access the applications.

Place the DVD in your DVD drive and wait. A program called *Multimedia Guide to the Periodic Table Launcher* should start automatically (Figure 1). If the DVD does not start automatically you can start it manually by opening the program called *autorun.exe*.

The purpose of the *Multimedia Guide* is described below. It includes a few documents that provide a more in-depth discussion of some of the underlying science. *Adobe Reader* is required to read these documents.

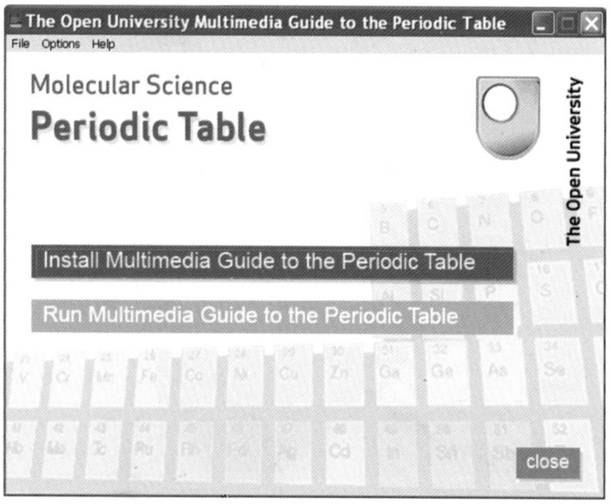

Figure 1 The *Launcher*.

During the installation you are offered several choices. We recommend that you accept the default suggestions where possible.

Once you have completed the installation you can run the *Multimedia Guide* (Figure 2).

Accessing the DVD contents

All components are accessed through the *Multimedia Guide* (Figure 2).

Select an entry in the left-hand panel and read the initial description of it in the right-hand panel. When you are ready to start the activity click on the 'Start' button.

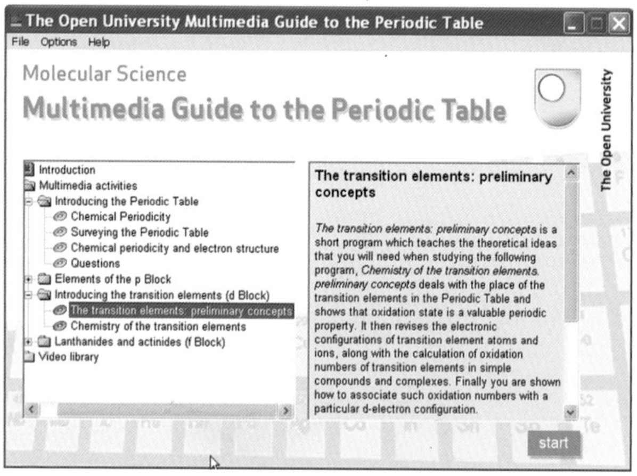

Figure 2 *The Multimedia Guide.*

Progress records associated with the multimedia activities

For some multimedia activities the software keeps track of your progress and stores information on the hard disk of your computer.

This facility allows you to exit from a program, switch your computer off and then resume again later. If you wish to delete the records stored on a particular activity you can do this from the activity's menu item called 'Erase progress records'.

Uninstalling the Periodic Table DVD

You can remove most of the supporting software from the hard disk of your computer by using the 'Add or Remove Programs' feature in the 'Control Panel' folder, which you will find on the 'Start' button under 'Settings'.